Light-Sport Gyroplanes

An introductory guide for discovering these unique aircraft

Ira McComic

AMIRADO PUBLISHING

LSG4-067

Amirado Publishing
2128 Heather Hill Lane
Plano, Texas 75075
www.AmiradoPublishing.com

ISBN 978-0-9887574-3-1

While every effort has been made to ensure the accuracy of the information contained in this publication, the author and publisher disclaims all responsibility for omissions and errors.

Issues discussed in this book and its associated website that relate to Federal Aviation Administration (FAA) regulations and practices are not legally authoritative. FAA regulations, orders, policies, and other documents are subject to change and interpretation. For complete and authoritative information about FAA regulations, orders, policies, and other documents, consult those publications or contact the FAA directly.

Light-Sport Gyroplanes

An introductory guide for discovering these unique aircraft

Table of Contents

Acknowledgments

So many persons contributed to the development of this book, either directly or indirectly, that it isn't feasible to acknowledge them all in this limited space. I can only express my appreciation to them by offering examples of particular persons from whose expertise and enthusiasm I have drawn. I gratefully acknowledge

Dr. Bruce Charnov for penning *From Autogiro to Gyroplane*, an authoritative history of gyroplanes;

Ron Menzie for enabling so many persons to become gyroplane pilots;

Greg Gremminger for his efforts in representing the gyroplane industry as the chair of the ASTM committee for light-sport gyroplanes and who continues to advance gyroplanes;

Dayton Dabbs for being a professional example of a new gyroplane generation and for embracing new challenges;

Desmon Butts for expanding the reach of gyroplane training;

Abid Farooqui for his timely response to my requests for information and for his insights on ASTM requirements;

John Rountree for his innovative approaches to solving problems;

Mike Grosshans, editor of the Southwest Rotorcraft enewsletter, for diligently communicating the special attraction of gyroplanes;

Vance Breese for being a responsible and inspiring representative of gyroplane pilots.

Thanks also to the countless others who have helped me gain a greater understanding and appreciation of these unique aircraft, a task—although a pleasant one—that continues.

Close to a hundred photos are included throughout the book, including photos I took at different times, ones from gyroplane manufacturers or their representatives, and ones from other individuals. I especially thank Jon Stark for photos of the Air & Space 18A and the McCulloch J-2, Joe Reinhard for the photo of the Pegasus Mark III, and Chris Lord for the photo of the Calidus with a smiling shark face. And a big thanks goes to Jack Fleetwood, avid aviator adventurer and photographer, for allowing me to use a number of photos from his vast collection.

I am also grateful for the support of my family. To Gretchen, Jennifer, Charlotte, Matt, Allison, Miranda, Andy, Sally, Cole, Brayden, Emerson—and to Jeff whose spirit lives on—this book is lovingly dedicated.

Ira McComic

Introduction

Flying is both exciting and soothing. It's exciting to break the bounds holding you to the ground and soar skyward, the vista of the horizon expanding before you, stretching into the distance, beckoning you to follow. And it's soothing to smoothly sail above the earth in a heaven of sunshine and blue sky, held aloft in a stream of unseen winds.

Photo 1-1 An Autogyro MTO Sport light-sport gyroplane

Persons choose to fly for different reasons. For some it may be a career, even a part-time one. For others it may be a combination of business and pleasure. For whatever reason, at the heart of that choice is the enjoyment of flying. And it's that enjoyment that makes light-sport aircraft so appealing. They're intended for the fun of flying.

Flying light-sport aircraft is more about the journey than the destination. It's for persons who fly for pleasure and aren't in a particular hurry to get somewhere and get it over with. Because light-sport aircraft are relatively small, the pilot often feels more connected to the aircraft. Compared to larger aircraft, flying a light-sport aircraft tends to be more a matter of classic stick-and-rudder flying. There's more feeling of being a part of the aircraft rather than managing a cockpit.

The Federal Aviation Administration (FAA) regulations that, in 2004, ushered in light-sport aircraft has fostered a new niche within general aviation, one that has become a fast growing segment of that branch of aviation. Along with light-sport aircraft, the regulations introduced a new level of pilot license—Sport Pilot—that made it possible for persons to earn a pilot license more quickly and less expensively. Also, the Sport Pilot license has made it possible for more pilots to prolong the joy of flying.

"Pilot license" is a common term for an airman certificate authorizing a person to fly an aircraft.

You might ask, "What exactly is a light-sport aircraft?" Instead of an exact, but longer answer (which will come later), a simpler and shorter answer is this: a light-sport aircraft is a relatively simple aircraft that's intended for recreational flying. Its maximum weight is limited to 1,320 pounds (when operated to and from land); it's allowed to carry up to two persons (a pilot and a passenger); and it can have a maximum continuous cruise speed of up to 138 miles-per-hour (120 knots).

A light-sport aircraft may be flown with a Sport Pilot license. Compared to a Private Pilot license, a Sport Pilot license requires half the minimum number of flight training hours. Also, a Sport Pilot license doesn't require an accompanying airman medical certificate. In place of a medical certificate, a Sport Pilot may use a valid driver's license. That's a brief description of the Sport Pilot license. Another chapter delves more completely into the Sport Pilot license and its privileges.

There are different kinds of light-sport aircraft. They can be airplanes, gliders, powered parachutes, weight-shift-control, or lighter-than-air, but when it comes to the pure joy of flying, there's one particular light-sport aircraft that hasn't been discovered yet by many persons: the gyroplane. Gyroplanes have unique features that make them an enjoyable choice for the kind of flying that characterizes light-sport aircraft.

This book is about light-sport gyroplanes; that is, ones that can be flown with a Sport Pilot license.

Photo 1-2 A Magni M-16 light-sport gyroplane.

Who this book is for

This book was written for persons who are interested in gyroplanes, especially those who may not know a lot about them but are curious to discover more.

The number of persons who currently fly gyroplanes is small. Although this core of gyroplane enthusiasts may find this book interesting, it's intended more for the larger number of persons who haven't discovered these unique aircraft. That includes pilots who fly other kinds of aircraft. Even among pilots and other persons who profess to know something about gyroplanes, I've discovered that often their knowledge is limited, and what they do "know" about gyroplanes may be mistaken or outdated.

This book is written primarily for US readers. There are two reasons for this. First, because I'm a pilot in the US, that's the country where I have direct experience. The second reason is because of the unique regulations for gyroplane pilots in the US compared to some other parts of the world. There's good news and bad news associated with that. The good news is this: with a Sport Pilot license, it's easier to begin, and continue, flying than ever before, and because FAA regulations allow Sport Pilots to fly light-sport gyroplanes, the benefits of a Sport Pilot license extend to flying light-sport gyroplanes. The bad news is that these regulations contain a disparity whereby gyroplanes are treated less favorably than other light-sport aircraft in the US. That disparity is discussed later in more depth in another chapter.

Even if you're among those who know something about gyroplanes, you may be surprised to learn about developments that are making gyroplanes better and safer. Especially if what you know about gyroplanes is limited to US providers, you may be surprised to learn what's available in Europe and other parts of the world and making them increasingly popular worldwide. Some of these models of sporty, stable, professionally designed gyroplanes have made their way to the US. Manufacturers, from overseas and in the US, are poised to bring other models to the US market, awaiting a favorable resolution of the regulatory disparity affecting light-sport gyroplanes in the US.

In brief, if you're avidly interested in gyroplanes, or simply curious, this book is for you.

Why this book was written

I wrote this book to help persons discover the unique enjoyment of flying light-sport gyroplanes. With more knowledge and interest about gyroplanes, the more the sport can grow. But growth alone isn't the goal. It's important that greater growth be accompanied by increased safety and responsibility. A light-sport gyroplane is a fun choice for flying, but operating one safely and responsibly is essential.

The lack of knowledge about gyroplanes is understandable. Information about them is sparse, especially recent information. Although several authoritative books have been published about gyroplanes, few of them are recent. Some of the older books have a wealth of technical information and contain fundamental principles that are still relevant. Excellent for the time in which they were written, these older books reflect the autogyros of their time but don't address newer, more modern gyroplanes. Only a few books are new enough to address the most recent light-sport gyroplanes. These few books, although informative, tend to be limited to only one particular manufacturer's line of gyroplanes.

Unlike some other kinds of aircraft, there are no periodicals devoted exclusively to light-sport gyroplanes. *Rotorcraft*, the enewsletter for members of the Popular Rotorcraft Association, has news and information that's mostly about gyroplanes but also carries information about personal helicopters. Only a few other aviation periodicals provide any coverage at all of gyroplanes. Notably, *Powered Sport Flying* magazine has a section devoted to gyroplanes that's refreshingly current.

A number of websites have some information about gyroplanes, yet many of the websites, especially those that only touch on gyroplanes, sometimes contain inaccurate or incomplete information. For the few websites that focus on gyroplanes exclusively, some of them are kept current. Others, though, are updated seldom and irregularly. Even websites belonging to some gyroplane providers, especially providers of older US models, go without updates for months and sometimes for years.

In another chapter of this book, you'll find a list of books, periodicals, websites, and other information resources about gyroplanes.

Although I've attempted to make this book as current as possible as of its publication, developments and events related to gyroplanes can happen faster than a book can keep up with them. That's why there's a companion website for this book with information about those developments and events.

For information about new developments, events, and other things related to light-sport gyroplanes, visit

www.LightSportGyroplaneS.com

It's rare when any book claims to answer every question or provide everything there is to know on any subject, and this book doesn't claim to do that for gyroplanes.

Specifically, this book isn't intended to be a technical tutorial. When learning about gyroplanes, it's important to gain a fundamental understanding of some of the basic concepts of these aircraft and this book attempts to offer that, but it isn't a technical treatise nor does it discuss technical concepts in great depth. Also, this isn't a book to teach you how to fly a gyroplane. There are other books designed for that purpose.

This book is intended to provide fundamental information about gyroplanes, light-sport gyroplanes in particular, that can help you decide if they're something you want to learn more about and possibly fly.

To further aid that decision, the next chapter explores some things that make gyroplanes special.

Gyroplanes are Special

A gyroplane has features similar to both a helicopter and an airplane, yet the combination of those features, plus its own unique attributes, make it a distinctive kind of aircraft.

Because a gyroplane has a rotor rather than wings, some persons mistake it for a helicopter at first glance. The primary difference between a gyroplane and a helicopter is the source of the force that turns the rotor. A helicopter's rotor is powered by the aircraft's engine. As a result of the engine's connection to the rotor, a helicopter's spinning blades produce the lifting force that holds the aircraft aloft, even when it's hovering stationary above a spot on the ground. A helicopter's rotor serves as both the aircraft's wings and propeller. Because a helicopter's engine powers the rotor, the rotor can propel the helicopter up—even straight up—forward, backwards, and sideways. To go down, a helicopter reduces its rotor's lifting force.

Similar to a helicopter, a gyroplane's spinning rotor produces the lifting force to hold it aloft. But in contrast with a helicopter, a gyroplane's engine doesn't power the rotor in flight. Instead, it's the air moving through the gyroplane's rotor that causes it to spin. So, how is the air made to move through the rotor? The same way air is made to move over an airplane's wings, by a propeller forcing the aircraft through the air.

Photo 2-1 An MTO Sport gyroplane from Autogyro Gmbh flying overhead

Like an airplane, a gyroplane has a propeller. Most airplanes, including single-engine light-sport airplanes, have their engine and propeller at the front and the propeller pulls the airplane through the air, causing air to flow over the wings. A propeller on the front of an aircraft is called a *tractor* configuration.

Some airplanes have a propeller at the back that pushes the airplane through the air. An aircraft with a propeller on the back is called a *pusher.* Although some light-sport gyroplanes have a propeller on the front, most of them are pushers.

Photo 2-2 This Little Wing gyroplane has a tractor configuration

Photo 2-3 This MTO Sport, like most light-sport gyroplanes, has a pusher prop

The purpose of an aircraft's engine is to provide the power to move the aircraft through the air. Air moving over an aircraft's lifting device, such as the wing of an airplane or the rotor of a helicopter, is what causes the lifting force that enables the aircraft to fly. Whether it's a helicopter, an airplane or a gyroplane, if the engine stops running, the aircraft comes down. If it's an airplane, the pilot can ease the nose down to allow the airplane to glide. Although the airplane is falling, it's also still moving forward through the air and that forward movement causes enough air to move over the wings to produce sufficient lift that the pilot can control the descent.

Likewise with a helicopter, when there's no engine power propelling the rotor, the helicopter starts falling. Yet, there's air still moving through the rotor, enough so that the rotor continues to spin and produce enough lift—similar to the effect of the air flowing over the wings of a gliding airplane—to allow the pilot to control the descent. When a helicopter's engine isn't powering the rotor and instead the rotor is propelled by air moving through it, the rotor is said to be in *autorotation*.

In flight, a gyroplane's spinning rotor is constantly in a state of autorotation. The engine, turning the propeller, moves the gyroplane through the air, and the air moving through the rotor keeps the aircraft aloft. It's similar to an airplane's propeller moving the airplane's wings through the air. Should a gyroplane's engine stop, the gyroplane's rotor continues in autorotation, allowing the pilot to control the gyroplane's descent similar to a helicopter in autorotation or to an airplane in a power-off glide.

Even though a gyroplane may appear to be a hybrid between a helicopter and an airplane, it's a distinct kind of aircraft with its own unique handling and performance characteristics. It's also uniquely fun to fly.

> *I have people come to me [for training] from different backgrounds. People who have never been involved in aviation see a gyroplane and are intrigued by it. I have a fifteen-year-old student who's getting ready to solo when he turns sixteen. Experienced pilots also see gyroplanes and say, "that looks like so much fun." A guy who owns a famous flight museum in Florida—he has owned over a hundred and fifty airplanes—came to fly with me and he absolutely loved it. I've had helicopter pilots who flew in Afghanistan, Black Hawk pilots from the Army, and a V-22 Osprey pilot all wanting to learn to fly gyroplanes.*
> *— Dayton Dabbs, gyroplane CFI*

Hands-down, a light-sport gyroplane is easier to fly than a helicopter. Compared to a helicopter, a light-sport gyroplane has a simpler rotor system. For one thing, there's no collective lever as there is with a helicopter. And because a gyroplane's rotor isn't powered in flight, there's no need for a tail rotor to counter the torque of the overhead rotor and no need for the pilot to manage anti-torque pedals.

A gyroplane is no more difficult to fly than an airplane. Some say it's easier. For example, unlike most airplanes, there aren't any wing flaps to fiddle with.

Photo 2-4 Light-sport gyroplane flight controls

A light-sport gyroplane has three fundamental flight controls:

a stick-like device called a *cyclic* for pitch and roll control,

rudder pedals similar to those on an airplane for yaw control,

and a throttle for power control.

What is it that makes a light-sport gyroplane special compared to other light-sport aircraft? Perhaps the most obvious thing that's special about light-sport gyroplanes is that they have rotating wings rather than fixed wings. Most other light-sport aircraft—airplanes, gliders, weight-shift control trikes, and powered parachutes—have fixed wings, even though the wing of a powered parachute must be inflated by ram air. The other kinds of light-sport aircraft—balloons and airships, don't have wings at all, yet even a hot air balloon's buoyancy, and that of an airship, is a result of an air envelope that's fixed.

Of all the different kinds of aircraft that may be categorized as a light-sport aircraft, only the gyroplane has rotating wings (collectively called a rotor). And it's a gyroplane's rotor that provides the aircraft with unique performance characteristics.

An airplane's wing is rigidly fixed to the airframe. With a gyroplane, though, its airframe is suspended underneath the rotor, which is attached to the airframe by a mast. Because the rotor can tilt in relation to the mast, there's a degree of independence between the disk formed by the plane of the spinning rotor and the gyroplane's airframe. An airplane pilot flies an airplane, but it may be said that a gyroplane pilot flies a rotor. The point is that a gyroplane pilot is distinctly concerned with rotor management, whereas an airplane pilot hardly ever thinks about managing the wings as distinct from the airframe.

Because a gyroplane has spinning wings, it has a unique property among light-sport aircraft: it's inherently stall resistant. Purists go so far as to state that a gyroplane can't stall. Because stalls are a common cause for aircraft accidents, a gyroplane is safer in this respect than aircraft that can stall.

Some gyroplanes can fly at more than a hundred miles an hour, approaching the top speeds of light-sport airplanes, yet it's the gyroplane's ability to fly safely at much slower airspeeds that make it special.

> *I fly airplanes and I fly powered parachutes, but I enjoy flying gyroplanes more. A gyroplane is a mix between the two because I can fly low and slow in a gyroplane or I can go a hundred miles an hour.*
> *—Bobby Martin, powered parachute and gyroplane pilot*

Compared to an airplane, a gyroplane is generally more controllable at slower airspeeds. Gyroplanes can't hover, but because they can fly at slower airspeeds, they may appear to do so in strong winds. The ability to fly at slower airspeeds allows gyroplanes to takeoff and land in short distances. Especially in strong headwinds, they're capable of zero-ground-roll landings. A pilot can take off and land a gyroplane on a shorter runway or in a smaller open field than with most light-sport airplanes. Also, having the ability to land in a shorter distance and in a smaller area is a particular advantage in the event of having to make an emergency landing.

When operated within their limits by an experienced pilot—who also operates within his limits—gyroplanes are capable of impressive maneuvers. Quick and responsive, their controllability at slow airspeeds gives them a tight turning radius, and they can make approaches to landing at steep angles.

Photo 2-5 A gyroplane is capable of impressive maneuvers

In a gyroplane, I am a bird, completely away from feeling like I'm flying in a vehicle. For me, it's a little like riding a motorcycle except I'm not afraid some little ol' lady is going to run over me.
— Gary Buster, airplane and gyroplane pilot

Gyroplanes are relatively easy to store and transport. This can be an important economic consideration considering hangar costs in some places. Some light-sport gyroplanes can fit in a home garage or in a commercial storage unit. A gyroplane's rotor blades, when aligned along the length of the fuselage, make a narrow profile.

Photo 2-6 Light-sport gyroplanes in storage

A gyroplane's rotor blades and hub bar are relatively easy to remove and replace, making it practical to transport a gyroplane on a trailer, or possibly in the bed of a truck, to and from an airport.

Photo 2-7 Many light-sport gyroplanes can be transported on a trailer

Because of their short field takeoff and landing ability, it's feasible for some gyroplane owners to have a runway on their own property.

Photo 2-8 A light-sport gyroplane uses a grass runway

Due to the inertia of a gyroplane's spinning rotor blades, a gyroplane is more wind tolerant than most other light-sport aircraft. On days that many other light-sport aircraft are grounded because of wind speeds, a gyroplane may not be. During the more windy times of the year, some light-sport aircraft are limited to flying only during the early morning or late afternoon hours when the winds may be calm enough, whereas a gyroplane pilot is more likely to be able to fly throughout a larger part of the day.

Photo 2-9 A Magni M-16 prepares to take off with a strong crosswind

The first time I saw a gyroplane was one time when I had gone to the airport to fly my powered parachute. I had arrived at first light while the winds were calm. About nine or ten was when the gyroplanes pilots showed up. By that time, the winds had kicked up enough that I couldn't fly, and I wondered how come they could.
— Bobby Martin, powered parachute and gyroplane pilot

Photo 2-10 The two-place Magni M-22 Voyager has an open cockpit

Although not unique to gyroplanes, many of them have open cockpits, which some persons enjoy, probably the same ones who enjoy riding in convertibles with the top down. Nearly all earlier-era gyrocopters had open cockpits, as do most of the more modern light-sport gyroplanes, but for those who prefer an enclosed cockpit, there's that choice also.

Photo 2-11 The two-place Magni M-24 Orion has an enclosed cockpit

The pilot's visibility in a gyroplane is remarkable. Below, the view is similar to that of a high-wing airplane, and looking above, a gyroplane pilot can see through the spinning rotor.

Whereas most airplanes, including light-sport airplanes, have tractor configurations, most light-sport gyroplanes are pushers. There are relative advantages and disadvantages of both configurations, but one of the advantages of the pusher configuration typical of light-sport gyroplanes is better forward visibility. Pilots and other occupants don't have an engine and a prop in the way of their vision.

Photo 2-12 A single-place Aurora model Butterfly gyroplane with a pusher prop

Especially in an open cockpit aircraft, having the propeller in the rear can reduce the possible danger to arms and legs and, as Cousin Seymore might add, "other useful appendages to which folks are attached".

Prices for gyroplanes extend over a wide range, but are generally in the low to mid-range for light-sport aircraft. Especially if you calculate the amortized cost of a gyroplane, factoring in the number of additional hours you're able to fly it compared to other light-sport aircraft, a gyroplane can be more of a bargain. The fact is, however, that aviation isn't the least expensive sport a person could pursue. That may not be a consideration for some persons for whom money is no object. For the rest of us, though, who are mindful of budgets, a gyroplane can be relatively affordable.

> *Gyroplanes are a less expensive sport, but it's not a cheap sport. Aviation isn't a cheap sport. For gyroplanes, like other aircraft, there's a level of safety that can't be under priced.*
> *— Dayton Dabbs, gyroplane CFI and Magni dealer*

Gyroplanes are special also because of their novelty. Compared to most other kinds of aircraft, even other light-sport aircraft, gyroplanes are relatively rare. At least, that's true currently. They're becoming more common as they grow in popularity. Wherever they go, gyroplanes attract attention, even if it's due to curiosity.

Photo 2-13 Light-sport gyroplanes attract attention

I once had occasion to trailer a gyroplane behind an SUV for several hundred miles. Every time I stopped for fuel or other necessities, the gyroplane would attract persons to it.

Hardly any of them knew much about gyroplanes, but as I chatted with them about gyroplanes and about flying them, I could see the eyes of many of them light up. It may have been my imagination, but I thought sometimes I saw some envy there too. In any case, I felt almost like a celebrity.

> *I became interested in gyroplanes when I was fourteen years old. I saw the James Bond movie* You Only Live Twice *[with the gyroplane in it] and I thought that was the coolest thing I'd ever seen. I said, "One of these days, I'm going to get a gyroplane."*
> *— Chauncy Surry, gyroplane owner and pilot*

At the heart of choosing a favorite kind of aircraft to fly, as with choosing other pursuits, it comes down to a matter of personal preference. For example, my brief experience flying gliders and my even briefer experience with hot air balloons taught me two things: 1) an admiration for pilots who have the finesse to fly things without engines, and 2) that's something I didn't want to do. For me, there's comfort in having an engine to compensate for unexpected changes in wind and other elements and in unintended changes in altitude and airspeed.

I enjoy flying airplanes, but I enjoy flying rotorcraft more. To simplify the distinction, airplanes are for high and fast flying; rotorcraft are for low and slow flying. Rotorcraft includes helicopters as well as gyroplanes, but between those two, gyroplanes are more accessible because they can be flown with a Sport Pilot license and helicopters can't.

> *I love to fly powered parachutes and I love to fly airplanes, but I love to fly gyroplanes more.*
> *— Bobby Martin*

I'm a commercially licensed pilot and I understand the financial importance of payloads and precise ETAs, but the profit from flying light-sport aircraft is measured in fun rather than in dollars and deadlines. According to that philosophy, if you're interested in flying light-sport aircraft—that is, if you want to fly for fun—consider gyroplanes.

How safe are gyroplanes?

In helping decide if gyroplanes are something you want to fly, let's confront a question head-on: how safe are gyroplanes? The short answer is they're as safe as anyone chooses to make them and fly them.

Doug Barker, the president of the Popular Rotorcraft Association, has this to say about the risk of flying in general and the relative safety of flying gyroplanes

> *Flying is inherently more risky that some other activities. We should go into this sport with our eyes wide open, knowing the risks and doing all we can to manage them to an acceptable level. Done with the right attitude, flying gyroplanes can be as safe as many of the other activities some people choose to pursue.*
> *— Doug Baker, PRA president*

As a pilot licensed to fly airplanes and helicopters, as well as gyroplanes, I enjoy talking with pilots of all kinds. Many of these conversations have led me to new discoveries about flying and an appreciation for different perspectives about aviation. Most pilots, I've found, are knowledgeable and want to stay informed about nearly all aspects of aviation. But when I mention "gyroplanes" to persons, pilots included, I often get a blank look. On the other hand, if I use the term "gyrocopter", it more often strikes a familiar note and often a sour one. "Oh, yeah," I might hear, "you mean those flying lawn chairs that are always crashing and killing people?" Based upon limited knowledge, they're certain that gyroplanes are something they would never want to fly. This opinion, shared by a number of pilots, is the same one sometimes held by non-pilots. And, in part, I agree with them. There are some gyrocopters I wouldn't fly.

I won't deny that gyrocopters have a reputation for being unsafe. And it's not an entirely unfair reputation. Gyrocopters have a dubious past, one too often storied with unsafe designs, unsafe construction, unsafe practices (especially self-training), and unsafe attitudes.

That's ironic, given that the fundamental impetus for the gyroplane was to improve the safety of flying. The autogyro, as the gyroplane was first called, had its invention in the concept of improving safety: to remedy airplane stalls. The first successful model of an autogyro was essentially an airplane modified with the addition of a rotor. And it accomplished the purpose for which it was designed; it provided a way for an airplane to continue flying even after its wings stalled. With further insights and developments, it was discovered that an aircraft with a rotor offered other advantages, such as more maneuverability at slower speeds as well as shorter takeoff and landing rolls, a particularly useful safety feature in the event of an engine failure. The implementation of those features led to commercial gyroplanes that proved safe in both design and operation.

Following the decline of commercial gyroplanes, it was afterwards, during the gyrocopter era, that the fundamentals of safe design, sound construction, and sensible operation developed during the eras of autogyros and type certificated gyroplanes were either forgotten, unlearned, or neglected by some amateur gyrocopter enthusiasts, predominantly those homebuilders whose emphasis was on affordability at the expense of safety.

I'm not an apologist for gyrocopters, yet I think gyroplanes deserve new consideration, especially in light of current developments.

Gyroplanes are now entering a new era. Discoveries made in the earliest era of gyroplanes—in the 1930s—and that positioned them as a safer alternative to airplanes, are being re-discovered.

Photo 2-14 This Magni M-16 is an example of modern gyroplanes of the light-sport gyroplane era

Fundamentals of safe design implemented by professional aircraft engineers and embodied in type certificated gyroplanes used for commercial operations are now being applied in the design and development of newer models of light-sport gyroplanes, ones that have the added advantage of being built with materials used in more modern aircraft. Gyroplanes of this new era are being designed with sound engineering principles and produced globally using industry standards similar to those of other aircraft. These gyroplanes of the light-sport era are more stable and better built than the often slapped-together gyrocopters of the intermediate gyrocopter era.

> *It's not a matter of gyroplanes not being safe anymore. Gyroplanes [coming out now] are every bit as safe as a [Cessna] 172 or any other training aircraft ..."*
> *— Dayton Dabbs, gyroplane CFI*

The new light-sport gyroplane era offers the prospect of better engineered, better built, and safer-to-fly gyroplanes. That's especially true in the parts of the world where those countries' regulations have fully embraced gyroplanes of this new era. In the US, these gyroplanes are also being adopted, but at a slower rate. This slower adoption in the US is largely due to an FAA regulatory disparity that prevents gyroplane manufacturers from competing equally with other light-sport aircraft.

Light-sport gyroplanes are poised to demonstrate their full potential, and when the FAA removes the disparity that discriminates against light-sport gyroplanes, I believe US pilots will embrace light-sport gyroplanes as fervently as their counterparts in other countries.

To better understand the origins of light-sport gyroplanes, the next chapter discusses the development of gyroplanes across three eras—from their earliest beginnings in the autogyro era, through the commercial type certificated gyroplane era, and through the gyrocopter era—up to the current light-sport gyroplane era.

Gyroplane Eras

The development of gyroplanes can be divided into four overlapping eras: autogyros, type certificated gyroplanes, gyrocopters, and light-sport gyroplanes.

The autogyro era

When Orville, the younger of the Wright brothers, lifted off a sandy slope in North Carolina and flew a fragile, motorized glider 120 feet, it was an event heralded around the world. The achievement marked the advent of aviation, not only in the US but globally. It has been a motivation, directly or indirectly, for all aviation developments that followed, including gyroplanes.

When Orville Wright made that historic flight in the US, it was a crisp December day in 1903. Juan de la Cierva, a Spanish lad, was eight years old.

Photo 3-1 Juan De La Cierva

Cierva took an early interest in aviation. He and his friends enthusiastically experimented with kites and models. Cierva decided aviation would be his career, and at the age of sixteen, enrolled at the Civil Engineering College of Madrid. While a student there, and along with friends, Cierva built the first airplane constructed in Spain.

A year after graduating from college, Cierva entered an aircraft design contest initiated by the Spanish government. Cierva's entry was a tri-motored bomber with a wingspan of eighty-two-feet. The airplane flew but the test pilot wasn't experienced with flying such a large airplane and stalled it. Although the pilot walked away from the resulting crash, it dashed Cierva's chances of winning the contest.

Cierva pondered the cause of the crash. Was there a way to prevent accidents from stalls?

Airplanes fly as a result of air flowing over their wings. To stay aloft, an airplane must have enough air flowing smoothly over its wings and producing enough force called lift, to overcome the gravity that's pulling down on the airplane. With too little air flowing over the wings—for example, when the airplane is slowed too much—the airplane quits flying, or as it is said, it stalls.

A more pure description of a stall is to say it occurs when the airplane's wings exceed a critical angle of attack. When the critical angle of attack is exceeded, the airflow over the wings is disrupted so much that the airplane stalls.

A scenario for a stall is when an airplane is flying relatively slowly, say right after takeoff or when making an approach for a landing. If the pilot pulls back the nose too far, further slowing the airplane, it can stall. A stall close to the ground can be especially lethal.

Cierva's solution to remedying stalls was to equip an airplane with rotating wings, slender blades that spun atop a vertical mast. Even when the airplane was slowed so much that there wasn't enough air flowing over the wings that were affixed to the fuselage to keep the airplane flying, its rotating wings—its rotor—would continue to spin, producing enough lift from the air flowing through them to prevent the airplane from falling abruptly. Even if the airplane's engine were to quit in flight, the air rushing up through the rotor would continue to produce lift and, even at a low altitude, would allow the airplane to descend slowly to the ground rather than crashing.

In concept, an aircraft rotor is similar to the spinning leaves attached to the seed of the *gyrocarpus americanus* whose common name is the Helicopter Tree.

The word adopted to describe an aircraft's rotor blades being propelled by air moving through them to produce lift, rather than an engine powering them, was *autorotation.*

Cierva trademarked the name *Autogiro* for the aircraft he developed. A more generic name for this kind of aircraft is *autogyro*. Eventually, the US Federal Aviation Agency, before it became the Federal Aviation Administration, would choose the name *gyroplane* for this kind of aircraft.

As is the case with many new inventions, an airplane with rotating wings wasn't perfect at first. Cierva experienced a number of problems while developing the autogyro. Among the most notable problems discovered when testing early models was that, upon taking off, they had an alarmingly consistent habit of rolling over on their side. After more extensive testing, Cierva discovered the cause of this rolling tendency: dissymmetry of lift.

When an airplane, whose wings are fixed to the airframe, is moving straight ahead and level, each wing moves through the air at the same speed and each wing produces an equal amount of lift. But when an aircraft with rotating blades is moving through the air, even straight and level, each blade at various times during its rotation is moving at different speeds in relation to the air. When a spinning blade reaches one side of the aircraft and is turning into the oncoming airflow, its speed in relation to the airflow is faster than a blade on the other side of the aircraft, which is moving in the same direction as the airflow and hence at a lesser speed in relation to the airflow.

As Cierva discovered, if the rotor blades are attached rigidly to its mast, and without any other compensation, the advancing blade (the one moving into the airflow), produces a greater amount of lift, and because it's attached rigidly to the mast, that blade tilts the mast, causing the whole aircraft to roll.

It was a perplexing problem, but one for which Cierva found a simple solution. Rather than attaching the rotor blades to the mast rigidly, the blades were attached to the mast on hinges that allowed each rotor blade to lift and lower during its rotation, the advancing blade lifting and the retreating blade lowering. The dynamics of allowing each blade to seek its own level in the airflow compensated for the unequal lift. With a solution to dissymmetry of lift, Cierva developed a model that flew successfully.

Following the first practical flight of that autogyro in January 1923, it made several more flights in the next few days, the longest of which was two-and-a-half miles. The average speed for the flight was less than forty miles per hour, dangerously slow for an airplane with only fixed wings, but safe for an autogyro.

Cierva continued to develop and refine his autogiros. His success in creating an airplane immune to stalls created widespread interest. In 1925, supported by investors, he moved his operations from Spain to England where he founded Cierva Autogiro Ltd.

Although Cierva didn't personally fly the earliest models of Autogiros, he did pilot later models. In 1928 he flew one across the English Channel, and in 1930 he flew one from England to Spain.

Photo 3-2 Cierva completes his first flight across the English Channel

With the increasing acceptance of his company's Autogiros, Cierva began licensing the technology and patent rights to companies in other countries, including Germany, Japan, and Russia. In the United States, the rights were licensed in 1929 to a company headed by Harold Pitcairn, the youngest son of a prominent Pittsburgh industrialist.

Photo 3-3
Harold Pitcairn

Pitcairn was only two years younger than Cierva, and the Wright Brothers had spawned his interest in aviation. Pitcairn eventually became a licensed pilot, the license signed by Orville Wright.

By 1928, Pitcairn had a successful airmail service business. He was an established airplane manufacturer, and had experimented with rotary-wing aircraft whose blades were propelled by compressed air.

Pitcairn had first met Cierva three years before and had followed the Spaniard's development of the autogyro. Pitcairn traveled to Spain in July of 1928 to meet with Cierva again. Recognizing Cierva's accomplishment, Pitcairn obtained the rights to manufacture aircraft incorporating Cierva's technology, and the two companies—Pitcairn's in the United States and Cierva's in England—agreed to a reciprocal licensing agreement of patents.

While in England, Pitcairn purchased a Cierva C.8 Autogiro and arranged to have it fitted with an American-built Wright J-5 engine. After it was shipped to Pittsburgh, it became the first autogyro to fly in the US.

Photo 3-4 A Pitcairn PCA-2

Pitcairn, as head of the newly formed Pitcairn-Cierva Autogiro Company (PCA), began building autogyros and making improvements. In 1931, the company's PCA-2 became the first autogyro to receive a type certificate from the US aviation agency, a necessity for manufacturing aircraft for commercial operations. The PCA-2 was especially notable because of its innovative ability for prerotation, using the aircraft's engine to start the rotor blades spinning before takeoff, thus shortening the takeoff roll.

As recognition for the achievement in developing the PCA-2, Pitcairn and his associates were awarded the prestigious Collier Trophy. President Herbert Hoover presented the award to them in a ceremony on the lawn of the White House. To dramatize the event, a PCA-2 landed on the lawn.

Pitcairn promoted his product with a passion. In the US, the company drew widespread attention to its PCA-2. In addition to landing on the White House lawn, the PCA-2 was adopted by the Detroit News for newsgathering and photography. The Champion spark plug company used a PCA-2 for promotions. "Miss Champion" as it was called, flew over six thousand miles in the Ford National Reliability Tour. The Beech-Nut food company, famous for its Beech-Nut chewing gum, acquired a PCA-2 and arranged for Amelia Earhart to fly it, making numerous appearances across the country. Many of those appearances came in conjunction with Beech-Nut's sponsorship of Earhart in 1931 to become the first person to fly across the country in an autogyro.

Photo 3-5 Thomas Edison, Pitcairn Chief Test Pilot James Ray, and a Pitcairn PCA-2

In attempting the first US transcontinental flight in an autogyro, Amelia Earhart was beaten to that record. John M. Miller, also flying a PCA-2, accomplished the feat nine days before Earhart completed it. However, Earhart had earlier set an altitude record in a PCA-2, climbing to 18, 415 feet.

Although the PCA-2 could fly up to 120 miles per hour, it was the autogyro's ability to fly slowly that made it unique compared to airplanes. Especially advantageous was its ability to land and takeoff in shorter distances than airplanes.

Early on, autogyros were fundamentally airplanes with a rotor added. They had wings like an airplane and depended upon those wings for much of their control. Even though the autogyro wouldn't stall, its wings' ailerons lost much of their control authority in slow flight when there was less air flowing over them. Later on, the development of the ability to tilt the rotors as they spun on the mast allowed an autogyro to be controlled completely without airplane wings, even in slow flight. No longer needing wings, an autogyro could land in even shorter distances.

By the mid-1930s, the autogyro was the nation's aviation darling in the US. The 1934 movie, *It Happened One Night*, starring Clark Gable and Claudette Colbert, had an autogyro landing at a society wedding. In 1935, Pitcairn was promoting its model AC-35, whose rotor blades could be folded lengthwise over the fuselage and driven on roads, painting a dream of an autogyro in every garage. In 1945, the song "I'll Buy That Dream" included the lyrics "A honeymoon in Cairo with a brand new autogyro". But by that time, the autogyro had been left behind, like a bride at the altar, by another rotary-wing attraction: the helicopter.

Two events in particular are notable in the decline of the autogyro's fortunes. One was the tragic death of Cierva at the end of 1936, a passenger aboard an ill-fated Dutch KLM airliner that crashed. Ironically, the crash resulted from the airplane stalling after taking off. With the death of the charismatic Cierva, much of the energy that drove the Cierva Autogiro Company was lost. The second event was the introduction of the Focke-Wulf FW 61 helicopter, which in 1938 dramatically demonstrated its hovering capability by doing so inside a Berlin sports stadium.

A helicopter can remain airborne and stationary over a spot, something a gyroplane can't do inherently. But that ability of a helicopter, made possible because the engine directly powers the rotor in flight, makes a helicopter more complex than a gyroplane, not only more complex to build and maintain, but also more complex to fly.

An autogyro's engine can assist in spinning up the aircraft's rotor blades before takeoff to reduce the length of the takeoff run, but the engine doesn't directly power the rotor in flight.

As Cierva had developed and refined autogyros, he was aware of the concurrent efforts to develop helicopters, but he considered autogyros more practical when weighed against the helicopter's greater costs and complexity. Cierva concluded that even without a helicopter's ability to hover, an autogyro's capability for short takeoffs and landings made it a sufficient alternative to airplanes. Adding to Cierva's conclusion was the development of an autogyro's ability to perform jump takeoffs. By spinning up its rotors before takeoff beyond the rotational speed needed for flight, an autogyro could spring directly up for several feet before leveling off and gaining forward speed. Thus, an autogyro's takeoff ground roll could be reduced to practically zero.

The Focke-Wulf helicopter that caught the world's attention in 1938, two years after Cierva's death, incorporated rotary wing technology the German company had licensed from Cierva's company.

Further improvements with helicopters led to an increased positive perception of them. Especially important to the helicopter's future, the military considered it essential that a rotorcraft have the ability to hover, and with a world war threatening, the fortune of helicopters rose while that of the autogyro fell.

Pitcairn's fortunes were tied to those of Cierva, and his company eventually quit manufacturing autogyros.

Another US company, the Kellctt Autogiro Corporation, held a license from Cierva to build autogyros. Although the US Army used a Kellett KD-1 autogryo briefly, the Kellett company stopped making autogyros in the late 1940s, switching to helicopters.

Photo 3-6 A Kellett KD-1 autogyro

The type certificated gyroplane era

By the end of World War II, the autogryo had become an artifact of a bygone era. The Pitcairn and Kellet autogyros marked the transition from the era of autogyros to the era of type-certificated gyroplanes.

A Type Certificate from the FAA is essential in allowing a company to manufacture completed aircraft that can be used for a broad range of commercial operations. Obtaining a Type Certificate is a rigorous, lengthy, and expensive process rigidly controlled by the FAA.

Once a manufacturer obtains a Type Certificate for an aircraft design and a subsequent Manufacturing Certificate, the individual aircraft that are produced are inspected after they are completed to ensure they conform to the design of the Type Certificate before being issued what is called a "Standard" airworthiness certificate.

Collectively, less than ten models of Pitcairn and Kellett autogyros received Type Certificates from the US agency that regulated aviation at that time, the predecessor to the Federal Aviation Administration. Today, a few Pitcairn PCA-2s are still preserved, all in museums, and only one is flyable. No Kelletts are still flyable.

More than a decade after the Pitcairns and Kellets, there was a renewed interest in producing type certificated gyroplanes. Between 1961 and 1971, three models of gyroplanes received Type Certificates: the Umbaugh U-18 (later modified and named the Air & Space 18A), the McCulloch J-2, and the Avian 2/180. All of them, along with the Pitcairns and Kelletts, ceased production decades ago.

Photo 3-7 An Air & Space 18A

Photo 3-8 A McCulloch J-2

Although pieces of 18As and J-2s lay scattered across the country and are occasionally discovered, less than half a dozen Air & Space 18As are still flyable and even a lesser number of McCulloch J-2s. As for the Avian 2/180, only one of them, renamed the Pegasus Mk III, is known to exist as a complete aircraft.

FAA regulations still provide for type certificated gyroplanes, but except for a handful in the hands of collectors, there are no longer any type certificated gyroplanes in the US.

Photo 3-9 A Pegasus Mk III gyroplane

The gyrocopter era

In the late 1950s another era began, one of smaller, cheaper, anybody-can-build-one gyrocopters. Ironically, these successors to the autogyro, once positioned as a luxury aircraft for the wealthy elite, were promoted as a means for flying on a shoestring budget. The person who fathered this new era, Igor Bensen, developed a line of simply designed Gyrocopters that could be homebuilt and easily assembled from cheaply fabricated parts. Originally developed as rotary-wing gliders, Bensen's Gyrocopters could be fitted with small, low-cost engines.

Igor Bensen trademarked the name *Gyrocopter* for his line of simple, cheaply built machines. The trade name became a popular nickname for all machines of this ilk, referred to generally as "gyrocopters".

Bensen was born in Russia in 1917. The upheaval of the Russian Revolution led his parents to relocate to Czechoslovakia. After beginning studies at the age of 17 in Belgium, Bensen migrated to the US where he earned a degree in mechanical engineering. After college, Bensen was offered a job with the company headed by helicopter designer Igor Sikorsky but was unable to accept it because of his alien status. Instead, Bensen took a position with General Electric.

After Bensen became an American citizen in 1934, General Electric assigned him to a project developing a helicopter with a jet-tipped-powered rotor. As part of its research, the company had acquired a Kellett autogryo and Bensen learned to fly it. Further sparking his interest in autorotation flight, Bensen discovered the US Air Force had in its possession two models of rotary-wing, manned kites used in World War II: the Fa-330 used for observation with German submarines and the Rotachute, which was used briefly by England to lift barrage cables over ships as protection from enemy airplanes.

Based upon Bensen's request, the Air Force loaned General Electric a Rotachute for study and with the stipulation that it not be flown. Bensen flew it anyway.

A notable feature of later model Rotachutes was a two-bladed rotor system connected to a common hub that teetered on the mast. A teetering hub was a simpler method of dealing with dissymmetry of lift than separately hinged rotors.

General Electric eventually developed a rotary-winged glider, the GE Gyro-Glider, but nothing became of it. For Bensen, though, it was another story.

After leaving General Electric and after two years with helicopter manufacturer Kaman Aircraft, Bensen went out on his own. His newly formed company, Bensen Aircraft Corporation, offered for sale a rotary-winged glider. The B-5 Gyro-Glider—with an aluminum frame, plywood fin and rudder, and a teetering rotor hub with two blades—could be towed behind a car.

Bensen made refinements to the B-5 and introduced other models of towed gliders. Bensen added an engine to the B-7 model and dubbed it the B-7M. The B-7M was followed by the B-8M, equipped with a seventy-two horsepower, two-stroke McCulloch engine built for military drones.

Photo 3-10 A modified Bensen B-8M

Introduced in 1957, the B-8M was sold as a kit of parts and plans. It was simple to build, required only cheap materials, and compared to other aircraft, the engine was inexpensive. Bensen advertised in publications like *Popular Mechanics* magazine. The ads advocated buyers to "teach yourself to fly." They could start by building the B-8 as a glider, learn to fly it towed behind a car or a boat, and add the engine later.

Bensen exhibited a classic trait of an entrepreneur: find a need and fill it. He tapped into a market composed of persons who wanted to fly simply and cheaply. For them, he touted just the thing: a cheap, build-it-yourself aircraft that was so easy to fly you could teach yourself.

Bensen, who would later receive a Doctor of Divinity degree, promoted his Gyrocopters with evangelical fervor. The Bensen Gyrocopter proved popular not only in the US but also worldwide.

In response to the groundswell of interest in Gyrocopters, Bensen formed the Popular Rotorcraft Association (PRA) in 1962.

Rotorcraft is a term that includes both helicopters and gyroplanes.

The PRA became a congregation of builders and flyers who shared their experience and enthusiasm. The non-profit organization provided members a monthly magazine and a yearly fly-in. The PRA also served as a means to further market Bensen's products. Many of the founding members were Bensen dealers.

Available with plans, the Gyrocopter led to modifications and other experimental designs based upon it, gyrocopters that didn't adhere strictly to Bensen's plans. Several of these designs led to persons selling plans and kits of their own. One of those persons who developed a derivative of the Bensen Gyrocopter that proved popular was Ken Brock.

In 1972, ten years after founding the PRA and establishing its headquarters in North Carolina, the state where the Wright Brothers had first flown, Bensen stepped down as the organization's president and took the title of president emeritus. The PRA, now headed by Ken Brock, moved its headquarters to California.

Derivatives of the seminal Bensen design abounded in the US and in other countries, including two-place gyrocopters and ones with an added range of engines. Consistent with the mindset of hot rodders, homebuilders bolted more powerful engines—motorcycle engines, boat motors, and automobile engines—onto the small gyrocopter frame. It was a time of extensive derivations and experimentation. Often, changes and additions were made without an adequate understanding of their impact on the stability and handling characteristics of the aircraft. Homebuilders became their own designers and test pilots, a combination conducive to disaster when those aircraft were operated outside their uncalculated, and thus unpredictable, limits. Even in the hands of an experienced pilot, some of the designs were uncontrollable. Mechanical failures, lack of structural integrity, and loss of control were common. This trial-and-error approach led to numerous accidents, injuries, and deaths.

Photo 3-11 A variety of gyrocopters

Worse, some of these shaky garage designs were sold to other enthusiasts who often were less experienced that the original builders who had the skills, or good fortune, to escape the dangers inherent in those designs and modifications. Unrecoverable events—for example, terminal pilot induced oscillations (PIOs) and buntovers took their toll.

Despite warnings from some of the more knowledgeable gyrocopter veterans about the potentially dangerous design flaws of some of the most popular gyrocopters—things like high propeller thrust lines combined with inadequate horizontal stabilizers—these models were defended by their makers and continued to sell well to hobbyists.

Compounding the potential for accidents, training didn't appear to be a high priority among some enthusiasts. Perhaps that was the result of the idea that gyrocopters were so easy to fly a person could learn on his own.

The role of ultralight aircraft

Homebuilt gyrocopters were a part of the wave of interest that attracted persons to the idea of crafting their own flying machine and that led to the formation of the Experimental Aircraft Association (EAA) in January 1953. Members of the PRA were of kindred spirit with those in the EAA; many of them belonged to both organizations.

In 1981, the EAA parented the EAA Ultralight Association. The term ultralight is used to recognize a flying craft with extremely limited capabilities compared to other aircraft. In fact, the FAA refers to ultralights as "vehicles" rather than aircraft.

As the name implies, an ultralight is essentially defined by its weight, which must be less that 254 pounds to avoid the more stringent regulations applied to aircraft.

The less-than-254-pound limit applies to ultralights with engines and that operate to and from land. An ultralight is allowed to weigh more if

it's rigged with floats for water takeoffs and landings or if it's equipped with certain safety devices. An ultralight that doesn't have an engine—a glider for example—can't weigh more than 155 pounds.

The FAA recognizes ultralights in Federal Aviation Regulations (FAR) Part 103, one of the briefest of federal regulations. Essentially, as long as an ultralight conforms to the weight and performance limits set by Part 103 and doesn't get in the way of aircraft when it's being flown, it's free of FAA oversight. An ultralight isn't required to be registered nor is its operator required to have a pilot license of any kind.

In addition to lightweight flying craft such as lighter-than-air balloons and gliders, flying machines such as powered parachutes, airplanes, and rotorcraft may qualify as ultralight vehicles if they conform to Part 103.

Photo 3-12 An ultralight gyrocraft

Ultralights satisfied the cravings of persons who dreamed of flying with the birds and, like birds, free of the necessity for a pilot license. Also, because flying an ultralight didn't require a pilot license, it didn't require an FAA medical certificate either.

Part 103 was put in place in 1982. From the beginning and as time went on, there was a persistent cry from ultralight flyers for the weight limit of ultralights to be increased. Doing so would allow them to fly craft of greater size and with more performance.

The FAA's answer to that cry was the creation of regulations that ushered in a new segment of general aviation in the US: light-sport aircraft. And, for gyroplanes, it represented a new era: that of the light-sport gyroplane.

Those regulations, however, come short of allowing light-sport gyroplanes the same opportunity for success as all other light-sport aircraft. The importance of the impact of this regulatory disparity that has singled out gyroplanes deserves elaboration, and to grasp this disparity more thoroughly, it's helpful to understand it within the context of the overall regulations that established the Light-Sport Aircraft category and the accompanying regulations that established the Sport Pilot license.

The next chapter looks at the regulations for Light-Sport aircraft. Other chapters look at the regulations for the Sport Pilot license and what it takes to become a gyroplane Sport Pilot.

Light-Sport Aircraft

In September of 2004, after a lengthy process of rulemaking, the FAA implemented regulations authorizing a new category of aircraft. The regulations were created to fill the large gap between ultralight flying craft, which are largely unregulated, and highly regulated type certificated aircraft. This new category was designated Light-Sport and the aircraft certificated in this category were referred to as Light-Sport Aircraft (LSA).

The FAA's designation of "Light-Sport Aircraft" suggests three essential aspects embodied in the regulations. The FAA considers light-sport aircraft to be "aircraft" rather than "vehicles" as it terms ultralight craft. The word "light" in the definition implies, though, that the FAA limits the size of these aircraft, specifically as measured in weight. Though they can weigh more than ultralights, light-sport aircraft are among the featherweights of general aviation aircraft. The word "sport" in the name suggests a light-sport aircraft is intended primarily for fun; its operation for commercial purposes is limited.

Along with regulations for light-sport aircraft, accompanying regulations created a new level of airman certificate designed specifically to allow persons to fly these aircraft. This new level of pilot license was called Sport Pilot (SP). Collectively, these regulations are referred to as the LSA/SP regulations.

The LSA/SP regulations were the culmination of a lengthy effort to make recreational aircraft, and flying them, more accessible and more affordable. There were a number of groundbreaking innovations in the regulations. For example, the Sport Pilot license was the first pilot license authorized by the FAA that didn't require an accompanying aviation medical certificate. Also, these were the first regulations permitting aircraft to be manufactured in accordance with industry consensus standards rather than with the more extensive process required for type certificated aircraft.

Because light-sport aircraft were intended for sport and recreation flying, rather than the more extensive commercial possibilities for type-certificated aircraft, the FAA regulations allowed light-sport aircraft to be produced using standards developed by industry representatives. Those industry consensus standards, agreed upon and accepted by the FAA, could be used as the necessary measurement of safety required for light-sport aircraft. Although the process required for a light-sport aircraft manufacturer to meet those industry consensus standards is lengthy and expensive, it is far less so than for type certificated aircraft. It also requires less work on the part of the FAA.

Attesting to the relative simplicity of regulations allowing the use of industry consensus standards for LSA, the first manufactured LSA arrived on the US market less than a year after the LSA/SP regulations were implemented.

Industry consensus standards are composed of a body of individual standards. For example, each kind of light-sport aircraft has its own consensus standard for design and performance. Other standards include ones for materials, manufacturing processes, and on-going provisions for safety-of-flight issues.

Industry consensus standards are developed by a collaborative process between industry representatives and in which FAA representatives participate. ASTM International (a respected organization formerly called the American Society for Testing and Materials) was chosen to facilitate the process for developing those standards. The resulting standards must be accepted by the FAA before they may be used by a manufacturer.

According to the FAA, "the use of the consensus standard process assures government and industry discussion and agreement on appropriate standards for the required level of safety."

Once a body of consensus standards is accepted, the FAA requires LSA manufacturers to certify that the aircraft they produce meet those standards before the agency awards those aircraft an airworthiness certificate.

When the FAA introduced the LSA/SP regulations, the agency identified four aspects associated with light-sport aircraft. The FAA expressed it this way: "the FAA is creating a new rule for the manufacture, certification, operation, and maintenance of light-sport aircraft." From that perspective, let's discuss these four aspects, beginning with manufacturing.

SLSA manufacturing

Light-Sport Aircraft are designed and manufactured in conformance with industry consensus standards. Not only is the manufacturing process done in accordance with an industry consensus standard, the aircraft that are manufactured are based upon a prototype aircraft whose design and performance conforms to the industry consensus standard for that kind of aircraft.

As with other consensus standards, the design and performance standard for an aircraft is developed by a committee of industry representatives and on which the FAA is represented. Subcommittees (often termed task forces) were formed for each kind of aircraft that was allowed to be a light-sport aircraft (e.g., airplane, gyroplane, glider, etc.). Each subcommittee was chartered to develop the design and performance consensus standard for the particular kind of aircraft it represented. Overall, each design and performance standard addresses fundamentally the same issues; however, each standard is adapted for the unique nature of the kind of aircraft for which it was developed.

For an LSA manufacturer, the use of industry consensus standards means that, rather than designing an aircraft under the scrutiny of the FAA, the manufacturer directly conducts the design and testing of a prototype aircraft, including flight testing, to confirm that the prototype meets the industry consensus standard for that kind of aircraft.

When the prototype is proven to conform to that standard, the manufacturer certifies to the FAA that it does. At that point, other consensus standards require that the manufactured replicates of that prototype be built consistent with the prototype to ensure that the replicates also conform to the industry consensus standard for design and performance.

The consensus standards for manufacturing require a production process that ensures the material, the fabrication, and the assembly of the replicates are all consistent.

Yet other consensus standards require that any safety-of-flight issues that are discovered after the manufactured aircraft have been put into operation must be dealt with consistently.

The entire process is subject to audit by the FAA to ensure a manufacturer is conforming to all the consensus standards applicable for that manufacturing.

SLSA certification

Two of the fundamental documents required for all aircraft are a registration certificate and an airworthiness certificate.

An aircraft registration certificate is similar to a vehicle registration document. Among other things, it identifies the owner of the aircraft, and like a vehicle's registration document associates its owner with a license tag number, an aircraft registration certificate associates its owner with an N-number.

An N-number is the collection of numbers and letters that appear on the outside of an aircraft. It's called a "number" even though it may have some letters also. It's called an N-number because the numbers and possible letters are preceded by the letter N, which by international agreement identifies the aircraft as being registered in the US.

An N-number is easy to acquire. It only requires making a simple application and paying a small fee to the FAA requesting a number. An N-number can be reserved even before an aircraft is built.

An airworthiness certificate is similar to a vehicle's inspection sticker. An airworthiness certificate is issued following an inspection of a completed aircraft. The inspection assesses the aircraft's compliance with regulations for its intended category and its suitability for flight in relation to the purpose of its operation within that category. If the aircraft satisfies that assessment, it's granted an airworthiness certificate for that category and for that purpose. An airworthiness certificate is the document that grants authority for an aircraft to be flown. It may include specific limitations for the aircraft's operation.

As with other aircraft, individual light-sport aircraft are required to be registered and issued an airworthiness certificate. Factory-built Light-Sport Aircraft are issued "Special" airworthiness certificates. Compared to a Standard airworthiness certificate typically issued to type certificated aircraft, a Special airworthiness certificate more strictly limits the use of a Light-Sport Aircraft for commercial purposes.

There are different categories of a Special airworthiness certificate according to the aircraft's purpose. For LSA built by a manufacturer in compliance with industry consensus standards, the category is designated Light-Sport and is for the purpose of Operating as a Light-Sport Aircraft. A Special airworthiness certificate in that category is termed SLSA for short. By association, aircraft issued an SLSA airworthiness certificate are called SLSA.

Light-Sport Aircraft built by a manufacturer, ones that are qualified to receive a Special airworthiness certificate in the Light-Sport category, are referred to as SLSA.

Although aircraft certificated with a Special airworthiness certificate are limited in regard to commercial operations, SLSA aircraft are permitted to be rented, including rental for flight training.

In conjunction with factory-built LSA (that is, SLSA), the FAA also allows an SLSA manufacturer to provide the aircraft in kit form, one whose parts are produced by the manufacturer and whose assembly can be made by it builders. A kit-built LSA is eligible for a Special airworthiness certificate, but one of a different category: Experimental.

Experimental is one of several categories of Special airworthiness certificates and the Experimental category itself has several different designations, depending upon the aircraft's purpose. An LSA assembled from a kit conforming to a factory-built LSA and completed in exact compliance with the directions of the manufacturer, is issued a Special airworthiness certificate in the Experimental category and with the purpose of Operating as a Light-Sport Aircraft.

An aircraft produced from a kit conforming to an SLSA parent is referred to as an ELSA.

Before a manufacturer may offer an ELSA version of an aircraft, it must first have manufactured a fully factory-built SLSA. Although an SLSA manufacturer isn't required to offer a kit-built version of an aircraft, an SLSA manufacturer cannot offer ELSA offspring unless it has produced at least one SLSA parent.

By current FAA regulations, there can be no ELSA without an SLSA parent. When the LSA regulations were first implemented, however, the FAA did allow some light-sport aircraft to be issued an Experimental Light-Sport Aircraft airworthiness certificate without a factory-built parent. These aircraft included two-place ultralights that had been previously exempted from some ultralight limits because they were used for training. This offer for an Experimental Light-Sport Aircraft airworthiness certificate was also offered as an enticement—some say amnesty—for owners of "fat" ultralights, ones that may have been factory-built and were being operated as ultralights even though they were actually too heavy to legally qualify as ultralights. For a period of time after the LSA regulations came into force, those ultralight owners could register their craft and apply for a special airworthiness certificate allowing those craft to be categorized as an ELSA. That period ended in 2008.

SLSA operation

Industry consensus standards for SLSA also include provision for ongoing safety-of-flight issues.

ASTM Standard F 2295, entitled "Standard Practice for Continued Operational Safety Monitoring of a Light Sport Aircraft", provides for the establishment of "a method by which safety of flight issues are discovered, evaluated, and corrected for the purpose of maintaining operational safety of an LSA."

This standard requires manufacturers to maintain an Operational Safety Monitoring System and assigns duties and responsibilities between manufacturers and owners/operators. It further requires manufacturers to evaluate safety-of-flight concerns using an Operational Safety Risk Assessment Procedure.

If corrective action is determined to be warranted, the manufacturer must issue a Notice of Corrective Action, categorized as either a Safety Alert, which requires immediate action; a Service Bulletin, which does not require immediate action, but does recommend further action; or a Notification, which need not necessarily recommend future action, but does provide "airworthiness information".

SLSA maintenance

The LSA/SP regulations allow a wider range of persons to perform maintenance of SLSA compared to type certificated aircraft. Maintenance, as well as inspection, of SLSA may be performed by a person holding an A&P rating. This is the rating generally required for persons who maintain and inspect type certificated aircraft. However, LSA regulations provide for maintenance and inspection to be performed by other persons, including owners of an LSA.

Except for minor issues, only certified mechanics are allowed to perform maintenance of a type certificated aircraft. Certified mechanics are persons holding a certificate with either an airframe or powerplant rating. Many certified mechanics hold both ratings, leading to the common shorthand reference for them as "A&Ps".

To become an A&P requires extensive education or supervised on-the-job experience. For an aspiring A&P taking the education route, it requires nearly two thousand hours of coursework at an aviation maintenance school approved by the FAA. Alternately, both ratings require a combined experience of two-and-half years.

In addition, A&Ps are required to have even more experience to perform annual inspections required for most aircraft. To do that, an A&P must be licensed for at least three years, among other requirements.

The point is, becoming an A&P isn't a trivial endeavor.

For SLSA, a person who satisfactorily completes an authorized course can legally perform the maintenance and inspection of any SLSA within the scope of that course, including his own aircraft. There is a specific course for each kind of SLSA; for example, airplane. The result of the completion of that course is a Light Sport Repairman with Maintenance Rating (LSRM) certificate. Obtaining a LSRM certificate is less stringent that obtaining an A&P rating. For example, a course leading to an LSRM certificate for airplanes totals 120 hours.

To perform maintenance on some components of SLSA may require more than an LSRM certificate. The SLSA manufacturer may prescribe additional qualifications or stipulations for maintenance. Also, some SLSA may have FAA certified parts that require more specific certification for persons performing maintenance on those parts.

For ELSA, as with other aircraft having an Experimental airworthiness certificate, a person doesn't need to have a certification to perform maintenance on that aircraft. However, not just anyone can perform the annual condition inspection.

An owner of an ELSA who satisfactorily completes a 16-hour course for the kind of light-sport aircraft he owns may be permitted to perform the annual condition inspection of his aircraft. To be authorized to perform that inspection after completing the course, the owner must make an application to the FAA for a certificate to perform the inspections. The authorization is allowed only for the owner of a specific ELSA; that is, for an aircraft with a specific N-number.

Definition of a light-sport aircraft

As part of the LSA/SP regulations, the definition of light-sport aircraft was added to Part 1 of the FAA regulations (the "Definitions" part of FAA regulations). The definition reads

> ***Light-sport aircraft*** **means an aircraft, other than a helicopter or powered-lift that, since its original certification, has continued to meet the following:**
>
> **(1) A maximum takeoff weight of not more than—**
>
> **(i) 1,320 pounds (600 kilograms) for aircraft not intended for operation on water; or**
>
> **(ii) 1,430 pounds (650 kilograms) for an aircraft intended for operation on water.**
>
> **(2) A maximum airspeed in level flight with maximum continuous power (V_H) of not more than 120 knots CAS under standard atmospheric conditions at sea level.**
>
> **(3) A maximum never-exceed speed (V_{NE}) of not more than 120 knots CAS for a glider.**
>
> **(4) A maximum stalling speed or minimum steady flight speed without the use of lift-enhancing devices (V_{S1}) of not more than 45 knots CAS at the aircraft's maximum certificated takeoff weight and most critical center of gravity.**
>
> **(5) A maximum seating capacity of no more than two persons, including the pilot.**
>
> **(6) A single, reciprocating engine, if powered.**
>
> **(7) A fixed or ground-adjustable propeller if a powered aircraft other than a powered glider.**
>
> **(8) A fixed or feathering propeller system if a powered glider.**

(9) A fixed-pitch, semi-rigid, teetering, two-blade rotor system, if a gyroplane.

(10) A nonpressurized cabin, if equipped with a cabin.

(11) Fixed landing gear, except for an aircraft intended for operation on water or a glider.

(12) Fixed or retractable landing gear, or a hull, for an aircraft intended for operation on water.

(13) Fixed or retractable landing gear for a glider.

The definition of light-sport aircraft in FAA Part 1 defines the nature of aircraft that may be certificated with an SLSA airworthiness certificate. In addition, it defines the nature of aircraft that may be flown by Sport Pilots. In other words, for an aircraft to be eligible to be flown by a person with a Sport Pilot license, that aircraft must satisfy the definition in Part 1, and if it does, it may be flown by a qualified Sport Pilot regardless of how the aircraft is certificated.

Sport Pilots may fly SLSA aircraft; that is, aircraft certificated with a Special airworthiness certificate in the Light-Sport category. In addition, Sport Pilots may fly other aircraft that aren't certificated as SLSA as long as those aircraft meet the definition in FAR Part 1. To fly any light-sport aircraft, a Sport Pilot must be qualified to fly that category/class of aircraft.

Within the Part 1 definition, several kinds of aircraft are eligible to qualify as a light-sport aircraft; other kinds can't. Helicopters and powered lift aircraft are specifically named as aircraft that aren't eligible to be defined as light-sport aircraft. Consequently, they can't be flown by Sport Pilots.

An example of a powered lift aircraft is the tiltrotor V-22 Osprey.

There are also stipulations in the definition that eliminate other, more specific kinds of aircraft from being flown by Sport Pilots. For example, although they aren't called out specifically, multi-engine airplanes aren't eligible to be light-sport aircraft because provision (6) of the definition defines a light-sport aircraft as having only one engine.

As another example, an aircraft that satisfies the weight limitation and performance limitations in Part 1 but has a capacity for seating more than two persons doesn't meet the definition of a light-sport aircraft. Even if no more than two persons actually occupy the aircraft when it's flying, it's still not defined to be a light-sport aircraft.

On the other hand, there are aircraft that may qualify as a light-sport aircraft that aren't directly specified in the definition. That's true even though the aircraft isn't specifically certificated as an SLSA. For example, a classic Piper J-3 Cub satisfies the definition of a light-sport aircraft even though the Cub has a type certificate, as do most other aircraft larger than the Cub. The Cub was manufactured long before LSA regulations were dreamed of, but because the little yellow tail dragger fits the definition in Part 1, it qualifies as a light-sport aircraft.

Other kinds of aircraft may also qualify as a light-sport aircraft besides airplanes and besides aircraft that are type certificated. For example, a gyroplane with an Experimental Amateur-Built airworthiness certificate, if it fits the definition in Part 1, is considered to be a light-sport aircraft.

As noted earlier, any aircraft that satisfies the definition of a light-sport aircraft in Part 1 is eligible to be flown by Sport Pilots regardless of how it's certificated. Many aircraft that fit the definition of a light-sport aircraft are certificated as Experimental Amateur-Built (EAB).

It may be confusing, but an Experimental Amateur-Built airworthiness certificate (EAB) isn't the same as an Experimental Light Sport (ELSA) airworthiness certificate. Both are Special airworthiness certificates in the Experimental category, but they are issued for different purposes. An ELSA airworthiness certificate is issued for the purpose of "Operating light-sport aircraft" whereas an EAB airworthiness certificate is issued for the purpose of "Operating amateur-built aircraft".

From a practical viewpoint, an essential difference lies in who (besides a certified A&P mechanic) may perform the annual condition inspection of the aircraft. For example, any owner of an ELSA certificated aircraft, whether it's the owner who originally built it or a subsequent owner, may gain the authority to perform the annual condition inspection of that ELSA by successfully completing an approved course and making an application with the FAA. However, with an EAB certificated aircraft, the only person (other than an A&P) eligible for the authority to perform the annual condition inspection is the builder of the aircraft.

With this overview of light-sport aircraft, the next chapter takes a closer look specifically at light-sport gyroplanes.

Photo 4-1 The Halley Apollo AG-1 qualifies as a light-sport gyroplane

Light-Sport Gyroplanes

The introduction of the Light-Sport Aircraft/Sport Pilot regulations marked the advent of the light-sport gyroplane era. The inclusion of gyroplanes within the definition of light-sport aircraft in FAR Part 1 allowed Sport Pilots to fly them.

Only a very few gyroplanes from the autogyro era still remain, all confined to museums. A few gyroplanes from the type certificated era remain; none of them qualify as a light-sport aircraft and therefore can't be flown by Sport Pilots. Still left, however, are many aircraft representative of the gyrocopter era, virtually all of which qualify, per Part 1, as a light-sport gyroplane.

Since the advent of the light-sport gyroplane era, more modern gyroplanes have been introduced that represent the potential to significantly influence the future of light-sport gyroplanes, yet a regulatory disparity embedded in the LSA/SP regulations has stalled their influence, preventing light-sport gyroplanes from reaching their full potential.

To qualify as a light-sport aircraft, a gyroplane must satisfy the definition in FAR Part 1, including its listed provisions. Although some of those provisions apply to other kinds of aircraft, one applies to gyroplanes directly. That provision reads

> **(9) A fixed-pitch, semi-rigid, teetering, two-blade rotor system, if a gyroplane.**

Notably, that provision specifically places gyroplanes among light-sport aircraft. Furthermore, it positions them potentially to be certificated as fully as all other light-sport aircraft defined in Part 1.

A rotor is one of the five major components of a light-sport gyroplane. The other components are a powerplant, an airframe, landing gear, and tail surfaces.

Photo 5-1 The major components of a light-sport gyroplane

In addition to all the other applicable provisions in Part 1 that a gyroplane must meet to qualify as light-sport, provision (9) specifically identifies the kind of rotor system that a gyroplane must have to qualify as a light-sport gyroplane. "Rotor system" refers to the components directly related to a gyroplane's rotor, including its rotor blades, its hub (that grips the rotor blades and that sits atop the gyroplane's mast), and the controls for the rotor.

In keeping with the general spirit that a light-sport aircraft be relatively simple, a gyroplane with a fixed-pitch rotor system is simpler than one with a variable-pitch rotor system. More specifically, the requirement for a fixed-pitch rotor system has a practical implication; it means a light-sport gyroplane can't have a collective control. In rotorcraft talk, a "collective", is a type of control that's used to increase or decrease the pitch of all the rotor blades together (collectively). It's a control mechanism common with helicopters. With gyroplanes, a collective can be used to effect jump takeoffs.

With a jump takeoff, the rotor blades are placed into a flat pitch by a collective control, or something similar, while the aircraft is held in place on the ground and the rotor blades are spun up by a prerotator to a rotational speed greater than needed for flight, adding extra energy to the rotor. For takeoff, the collective blade pitch is increased while simultaneously allowing the gyroplane to move forward. The result is that the extra energy in the rotor causes the gyroplane to spring into the air as it begins moving forward. For a gyroplane to have a jump takeoff ability requires it to have a more complex rotor system, specifically one where the pitch of the rotors can be changed; hence, a gyroplane with jump takeoff ability doesn't qualify as a light-sport aircraft.

Most of the later gyroplanes of the autogyro era and the gyroplanes of the type certificated era had jump takeoff ability.

In the world of rotorcraft, there are three classic rotor systems: fully articulated, rigid, and semi-rigid. All of these designs represent different ways of dealing with the forces acting on a rotor system, and all of them involve the rotor blades, the rotorcraft's hub, and the rotor blades' connection to the hub.

In the case of a semi-rigid rotor system, the blades are attached rigidly to the hub, but the hub itself isn't mounted rigidly to the mast. The hub can tilt at the top of the mast. With a two-blade semi-rigid rotor system, the blades are attached to the hub opposite of each other, and as the hub tilts, the two blades teeter on the hub; as one goes up, the other goes down.

Most of the gyroplanes now flying that meet the definition in Part 1 for a light-sport gyroplane are holdovers from the gyrocopter era. They fit the definition of light-sport gyroplanes in FAR Part 1 even though they're based on designs that originated before the light-sport regulations were dreamed of. This isn't a coincidence however. The FAA specifically had these gyrocopters in mind when it crafted the definition in Part 1, thereby allowing Sport Pilots to fly them.

The Bensen Gyrocopters that originated in the 1950s marked the beginning of the gyrocopter era. Following their introduction and in accordance with the spirit of exploration implied in the name given to "experimental" aircraft, other homebuilders began modifying Bensen's fundamental design. Those modifications included the use of more powerful engines than the ones Bensen used. Some of these modifications, without other accompanying design changes, contributed to these gyrocopters' uncertain stability. Although stable enough for some flying conditions, these aircraft exhibited their instability when exposed to less than optimum conditions or when mishandled by unknowing, often untrained, pilots. Accidents followed, including fatal ones involving buntovers.

A buntover, sometimes called a forward tumble, is a sudden nose-down action that places the aircraft so far beyond the limits of its pitch axis that there's no recovery. A buntover is more likely to occur with a gyroplane that has a combination of a rotor lift line in front of the gyroplane's center of gravity (GC), a propeller thrust line above the CG, and an inadequate horizontal stabilizer. That combination may result in marginal stability when encountering forces acting on the aircraft in flight.

Power pushover (PPO) is a term sometimes used for buntover to emphasize the role of a high engine thrust line in a forward tumble.

A buntover is a terminal event sometimes resulting from pilot induced oscillation (PIO), a condition where a pilot attempts to correct an aircraft's nose-high or nose-low attitude and in doing so overcorrects. Because of a lag between the control input to correct the condition and the aircraft's response, the pilot may over control—first in one direction, then another, each time with the results exaggerating the effect—leading eventually to an extreme nose-low condition for which there's no recovery or else a condition that causes the rotor blades to strike the tail of the aircraft.

Photo 5-2 Illustration of propeller thrust line and rotor lift line

A propeller thrust line and a rotor lift line are illustrated in Photo 5-2. The gyrocopter in this photo is modeled after the classic B-8M Gyrocopter, a pusher gyrocopter with a wooden prop powered by a McCulloch engine.

In Photo 5-2, the yellow arrow illustrates the approximate propeller thrust line. The propeller thrust line of a pusher gyroplane is an imaginary line drawn from the center of the propeller's hub, perpendicular to the rotating plane of the propeller, and extending through the engine and beyond. Propeller thrust is a vector; that is, it's a force with both magnitude and direction. The magnitude of the propeller thrust varies with the engine rpm; as the engine turns the propeller faster, the propeller produces more thrust. Because a propeller is attached to an engine, either directly or with a gear reduction unit, the position and direction of the propeller thrust line is determined by the engine mounting. The resulting propeller thrust line is said to be high, low, or center depending on its relative proximity vertically to the gyroplane's CG. For example, a gyroplane whose propeller thrust line passes relatively high above the CG is said to have a high thrust line (HTL). A gyroplane whose propeller thrust line passes close to the CG is said to have centerline thrust (CLT).

As an approximate visual reference, imagine where the thrust line intersects the pilot's body when seated in the gyroplane. As a rule of thumb, a gyroplane with CLT would have a thrust line passing approximately through the pilot's belly button. Looks can be deceiving, though, when estimating how close a gyroplane is to having centerline thrust. Although the thrust line is easy enough to visualize, the gyroplane's CG isn't.

On a pusher gyroplane, a propeller is designed to push the aircraft through the air. A high propeller thrust line also tends to push the nose of the aircraft down, and the greater the magnitude of the thrust, the greater the force pushing it down.

The red line in Photo 5-2 illustrates the approximate rotor lift line, depicting the force normally lifting up on the aircraft during flight. Like propeller thrust, rotor lift is a vector with both magnitude and direction. When a gyroplane is in flight, because the rotor isn't attached rigidly to the mast, the rotor lift line may move to some degree in relation to the horizontal component of the aircraft's CG. The magnitude of the rotor lift; that is, the power of the force pulling up on the gyroplane, may also change during flight due to pilot actions and in response to the air in which the gyroplane is flying. For example, a pilot at the top of a steep climb who pushes the nose down abruptly can, by unloading the rotor, decrease the magnitude of the rotor lift. A severe downdraft can also unload the rotor.

For stability, it's important that the rotor lift line be behind the aircraft's CG. With a high propeller thrust line, which tends to push the nose down, some gyrocopter designs placed the rotor force line in front of the CG, which tends to pull the nose up, as a counter to the propeller thrust. With such a design, and without other design compensations, if the rotor were to become unloaded sufficiently, the propeller thrust could overpower the rotor lift causing the nose to be pushed too far over for recovery.

A factor common to pilot induced oscillation and buntover is an inadequate horizontal stabilizer. An adequate horizontal stabilizer helps dampen aircraft pitch excursions and can help counter other situations that can lead to buntovers. Some authorities go so far as to state that a gyroplane can't buntover with an adequate horizontal stabilizer. An adequate horizontal stabilizer involves a combination of the size of the stabilizer's surface area, its aerodynamic shape, its distance behind the gyroplane's CG, and its position in relation to the propeller's slipstream.

The properties constituting an adequate horizontal stabilizer are often simplified by the expression Big Tail, Way Back (BTWB).

Admittedly, this is a simplistic explanation of some of the forces affecting gyroplane stability. Aircraft design is a complex engineering endeavor that requires a thorough understanding of factors affecting stability and a knowledge of how to implement designs to harmonize the reaction to forces acting on an aircraft that are often in conflict.

Light-sport gyroplanes of the gyrocopter era

This section takes a look at several aircraft representative of the gyrocopter era. Most of these aircraft qualify as light-sport aircraft that can be flown with a Sport Pilot license.

In looking at them, you can see examples of how some gyrocopter designers have dealt with stability issues within the narrow scope of propeller thrust lines and tail structures.

Some persons consider empennage, a term for an aircraft's tail structure, to be classier than "tail feathers".

In Photo 5-3 (below) are four single-place gyrocopters with different horizontal and vertical stabilizers.

Photo 5-3 Four examples of single-place gyrocopters

Let's look at these four gyrocopters more closely individually, starting with the second one from the right.

In Photo 5-4 (below) is a closer view of the tail structure of the gyrocopter first pictured in Photo 5-2. This gyrocopter has a relatively small horizontal stabilizer attached to the rear of its airframe's keel and below the vertical stabilizer to which the rudder is hinged.

Photo 5-4 A closer view of a tail structure similar to a Bensen Gyrocopter

In Photo 5-3 (on the previous page), the yellow gyrocopter at the far right is another Bensen-like gyrocopter with different horizontal and vertical stabilizers. Photo 5-5 (below) is a closer view of this gyrocopter's tail structure.

Photo 5-5 Another Bensen-like gyrocopter

The horizontal and vertical stabilizers are shaped and implemented differently than that of the other Bensen-like gyrocopter in Photo 5-4, but their surfaces areas appear to be approximately the same. This yellow gyrocopter's vertical stabilizer serves as a rudder and on which the two-piece horizontal stabilizer swivels as a unit.

Referring to again to Photo 5-3, the third aircraft down the line (the red one) has a "T-Tail" arrangement for the horizontal and vertical stabilizers. This aircraft is pictured in Photo 5-6 (below).

Photo 5-6 A custom gyrocopter with a T-tail

Photo 5-7 A closer view of a T-tail.

The T-Tail places the horizontal stabilizer and a larger portion of the vertical stabilizer directly in the propeller's slipstream. In Photo 5-7, you can see that the two parts of the horizontal stabilizer, one on each side of the vertical stabilizer, are swept back.

Notice that this gyrocopter has a cross bar for the main wheels similar to that of the previous two Bensen-like gyrocopters. The longitudinal keel is also similar, except the part of the keel in front of the cross bar has been elevated, probably done in connection with elevating the seat, which included extending the nose wheel. Elevating the seat effectively moves the gyrocopter's CG upward, putting it more in line with the propeller thrust line.

Referring again to Photo 5-3 on page 39, the gyrocopter at the far end of the line in that photo has yet a different horizontal and vertical stabilizer arrangement. Photo 5-8 (below) is a closer view of this fourth gyrocopter.

Photo 5-8 A custom gyrocopter

Compared to the gyrocopter pictured in Photo 5-2, this gyrocopter appears to have a more powerful engine and a higher propeller thrust line. Notice the larger surface area of the horizontal stabilizer compared to that of the previously pictured Bensen-like gyrcopters. A larger horizontal stabilizer can reduce the greater potential for a power pushover due to a more powerful engine. Notice also this gyrocopter's horizontal stabilizer is configured almost midway up the vertical stabilizer, placing it closer to the propeller's slipstream.

The previous four examples are comparisons of gyrocopters built from different designs. To further illustrate ways of dealing with propeller thrust lines and tail structures, consider the next five examples of gyrocopters, all Air Commands.

Photo 5-9 (next page) is an example of an early model single-place Air Command "low rider", so called because the fuselage sits low to the ground.

A propeller thrust line is expressed as high, low, or center in relation to a gyroplane's CG. When adjusting that relationship, it's usually not a matter of moving the propeller thrust line, but of moving the CG. As weight is added to an aircraft, the CG moves in the direction of where that weight is added.

On the Air Command in Photo 5-9, the gyrocopter's engine is mounted as high as necessary to provide clearance for the propeller blades. The weight of the fuselage, the weight of the fuel in the seat tank, the weight of the nose wheel and main gear, and especially the weight of the pilot—all located below the propeller thrust line—combine to displace the vertical component of the aircraft's CG below the propeller thrust line, leading to what appears to be a relatively high thrust line.

Photo 5-9 A classic single-place Air Command "low rider"

Photo 5-10 (below) is another example of a single-place Air Command with a partially lifted fuselage. Notice the Bensen-like tail structure.

Photo 5-10 A single-place Air Command with a partially lifted fuselage

Photo 5-11 is an example of a single-place Air Command with a more fully raised fuselage. By using longer struts to raise the fuselage, the vertical component of the aircraft's CG can be brought nearer the propeller thrust line, causing the aircraft in this photo to appear to be close to CLT. Notice the horizontal stabilizer is part of the combined rudder and vertical stabilizer.

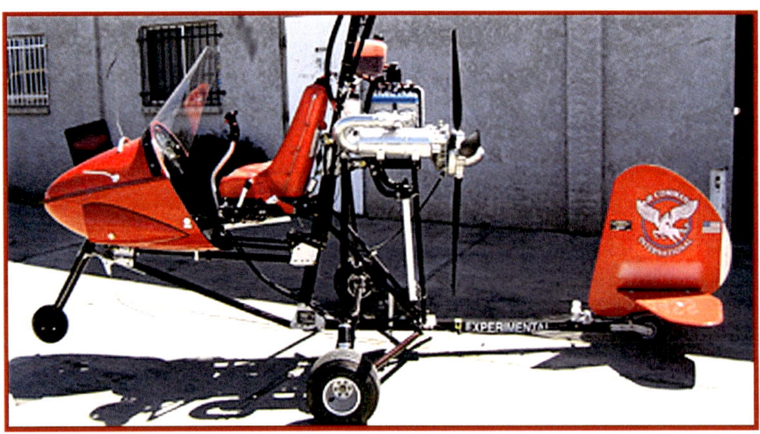

Photo 5-11 A single-place Air Command that appears to have CLT

Photo 5-12 is yet another single-place Air Command, this one also with a raised fuselage and that appears to have CLT. It also has a larger horizontal and vertical stabilizer. This gyrocopter has a four-cylinder, four-stroke engine that produces more power than the two-cylinder, two-stroke engine on the gyrocopter in Photo 5-11.

Photo 5-12 A single-place Air Command
with a tall vertical stabilizer and a horizontal stabilizer with upturned ends

In Photo 5-13 is a two-place Air Command. As with the single-place Air Command in Photo 5-12, it has a relatively powerful four stroke, four-cylinder engine and it has the same vertical stabilizer and horizontal stabilizer as the previous single-place Air Command.

Photo 5-13 A two-place Air Command with the same vertical stabilizer and upturned horizontal stabilizer shown in Photo 5-12

For a further comparison of tail structures, the following photos offer two examples of the Butterfly line of gyrocopters.

Photo 5-14 is a single-place Monarch model of the Butterfly line. Notice the horizontal stabilizer configured atop the vertical stabilizer. The bar protruding from the front of the rudder and near the top is weighted and positioned to help reduce the tail structure from fluttering.

Photo 5-14 A single-place Monarch Butterfly

Photo 5-15 is a two-place Golden Butterfly. The pilot sits in the front and the rear seat is elevated to allow the person there to see over the pilot's head. Notice this aircraft's version of a T-tail.

Photo 5-15 A two-place Golden Butterfly

Dominator gyrocopters from Rotor Flight Dynamics all feature a T-tail.

Photo 5-16 is a single-place Dominator with a high-mounted fuselage and a T-tail placed close to the propeller and in its slipstream. Notice the trim tab at the bottom of the rudder.

Photo 5-16 A single-place Dominator from Rotor Flight Dynamics

Photo 5-17 is a stretched, two-place version of the Dominator. Like the single-place, its fuselage is lifted and it has the same T-tail.

Photo 5-17 A two-place, tandem seating Dominator

All of the previous examples are ones of gyrocopters that originated in the US and are based upon designs that originated before the light-sport gyroplane era. Because of its impact in the US, one other example is worth noting even though it originated from another country.

Rotary Air Force (RAF) was a company originally based in Canada. The company's first product was a kit for a single-place gyroplane, the RAF 1000. It was followed by a two-place gyrocopter kit, the RAF 2000, which proved more popular. The company sold many of its two-place kits in the US and elsewhere.

Photo 5-18 An RAF 2000 GTX-SE

Although popular, the RAF 2000 became controversial due to a number of accidents involving it, many involving a loss of control, including buntovers. A particular focus of criticism for the RAF 2000's design was the gyrocopter's lack of a horizontal stabilizer.

As the number of accidents mounted, the company resisted suggestions that the gyrocopter's design was to blame, insisting the accidents were due to the lack of training of the pilots flying them.

Photo 5-18 is an example of a model of the RAF 2000 without a horizontal stabilizer. Notice also the gyrocopter's apparently relatively high propeller thrust line.

Some RAF 2000 kit builders, or their subsequent owners, added a horizontal stabilizer to their gyrocopter. Photo 5-19 is an example of an RAF 2000 to which a horizontal stabilizer has been added.

Photo 5-19 An RAF 2000 gyrocopter with a horizontal stabilizer

In 2003, a US company, Groen Brothers Aviation, developed a "Stability Augmentation Kit" for the RAF 2000. It included provisions for inverting the engine to bring the propeller thrust line closer to CLT as well as a prominent T-Tail. To market the stability augmentation kit, the company formed a subsidiary, American Autogyros, Inc. (AAI).

Photo 5-20 An RAF 2000 adapted with a stability augmentation kit

Subsequently, AAI developed its own two-place gyrocopter, which closely resembled the RAF 2000, and included AAI's stability augmentation kit.

Photo 5-21 An American Autogryos Sparrowhawk

Light-sport gyroplanes of the new generation

The examples in the previous section were those of gyrocopters designed prior to the LSA/SP regulations, which marked the beginning of the light-sport gyroplane era. Since that time, newer designs representative of the more modern gyroplanes of the light-sport gyroplane era have become available in the US.

Most of these more modern gyroplanes come from outside the US and originate in countries that have regulations similar to the US LSA/SP regulations. These gyroplanes are available in other countries as factory-built, ready-to-fly gyroplanes; however, because US regulations effectively don't allow factory-built light-sport gyroplanes*, these aircraft are available in the US in kit form. Below are examples of some of these light-sport gyroplanes.

***Specifically, because the FAA doesn't currently allow factory-built gyroplanes to be certificated as Special LSA (SLSA), it effectively prevents factory-built light-sport gyroplanes unless they are certificated as Experimental Exhibition, which places severe restrictions on how and where an aircraft may be operated, or if they are employed for public use, such as law enforcement. (See "The Future for Light-Sport Gyroplanes" for more about this issue.)**

Magni

Magni, an Italian gyroplane manufacturer, was among the first to provide gyroplanes in the US from abroad. Currently, Magni offers kit versions of three of its gyroplane models in the US: the M-16 Trainer, the M-22 Voyager, and the M-24 Orion.

Photo 5-23 Magni M-16 Trainer

Photo 5-24 Magni M-22 Voyager

Photo 5-25 Magni M-24 Orion

Photo 5-26 is a Magni M-16 Trainer. The Rotax 914 engine is standard with this model. Photo 5-27 is a Magni M-22 Voyager.

Photo 5-26 A Magni M-16 Trainer

Photo 5-27 A Magni M-22 Voyager

Like the M-16 Trainer, the M-22 Voyager is a two-place, tandem gyroplane, and like the M-16, the M-22 is powered by a Rotax 914 engine; however, with the M-22, controls in the back seat are optional. Although similar in many respects, the M-22 can be distinguished from the M-16 by the luggage pods on the side. In Photo 5-28 (below), the M-24's luggage pods are more prominent.

Photo 5-28 The Magni M-22 Voyager

Photo 5-29 is the Magni M-24 Orion, a two-place light-sport gyroplane with side-by-side seating. A cowling encloses the Rotax 914 engine.

Photo 5-29 The Magni M-24 Orion

Autogyro

Autogyro Gmbh, a German gyroplane manufacturer, offers kit versions of three of its gyroplane models in the US: the MTO Sport, the Calidus, and the Cavalon.

Pictured in Photo 5-30 (below) are two of its models. On the left and far right are the MTO Sport. In the middle is the Calidus.

Gyroplanes are being put to use in law enforcement in some parts of the US. As agencies deal with budget issues, some of them are discovering that gyroplanes can fulfill many of the roles performed by helicopters and at considerably less cost, typically about the cost of a couple of squad cars. Compared to a helicopter, the operating costs of a gyroplane are also considerably less. The US Department of Justice's Aviation Technology Program has helped local agencies with funding for gyroplanes.

Photo 5-30 Autogyro gyroplanes used in law enforcement

Photo 5-31 offers a closer look at an MTO Sport.

Photo 5-31 The Autogyro MTO Sport

In Photo 5-32 (below) are two Autogyro models together: The MTO Sport (left) and the Calidus. The two aircraft are similar. Both are two-place, tandem seating aircraft. The most obvious difference is that the MTO Sport has an open cockpit and the Calidus cockpit is enclosed. The Calidus has its engine enclosed in a cowling as well.

Photo 5-32 The open cockpit MTO Sport with the enclosed cockpit Calidus

Photo 5-33 The Autogyro Calidus

Photo 5-34 An Autogyro Cavalon

Photo 5-34 is the Cavalon model from Autogyro. It has an enclosed cockpit with side-by-side seating.

Halley

A relative newcomer to the US, the Apollo AG-1 comes from Halley, an established Hungarian manufacturer of Weight-Shift-Control aircraft (trikes).

Photo 5-35 A Halley Apollo AG-1 gyroplane

Photo 5-36 The Apollo AG-1 flying low level

Aviomania

Aviomania, headquartered in Greece, is one of the few companies from outside the US that offers a single-place gyroplane.

Photo 5-37 Aviomania single-place G1SA

The single-place Genesis G1Sa is pictured in Photo 5-37 (left). Most single-seat gyroplanes have seat tanks; that is, the pilot's seat doubles as a fuel tank. The Aviomania G1Sa's fuel tank is a separate component and is stationed below the seat.

Photo 5-38 (right) is the two-place Genesis G2Sa. Notice the saddlebag fuel tanks, one of which can be seen aside the passenger station in the rear.

Photo 5-38 Aviomania two-place Genesis G2SA

More information

For more information about these light-sport gyroplanes, and others, see the Directory in this book.

Other manufacturers are also establishing a presence in the US and are expected to have gyroplanes available in the US soon. More information about those developments can be found at

www.LightSportGyroplanes.com

The Sport Pilot License

To fly any aircraft, including a light-sport aircraft, requires a pilot license. Concurrent with the creation of the regulations for light-sport aircraft, the FAA added a new level of pilot license that persons could use to fly light-sport aircraft: the Sport Pilot (SP) license.

Two elements of a Sport Pilot license make it an especially good fit for some persons. First, for persons who don't already have a pilot license, a Sport Pilot license provides a possible avenue to a pilot license in less time and less expensively than for other levels of pilot licenses. A Sport Pilot license requires the least number of minimum hours of training than any other higher level of license. As an example, a Private Pilot license to fly an airplane requires at least forty hours of flight training. For a Sport Pilot license, it's half that. Because a Sport Pilot license can be achieved with a lesser number of training hours, it can be not only faster to attain, but also less expensive.

The FAA prescribes a minimum number of hours of training for each level of pilot license, but it doesn't specify a maximum. Most persons who earn a pilot license have more than the minimum hours of training required for that level of license.

The second element that makes a Sport Pilot license a good fit for some persons is that, unlike all higher levels of pilot licenses, a Sport Pilot license doesn't require an accompanying airman medical certificate. Instead, a Sport Pilot may use a valid driver's license. That's also why a Sport Pilot license can be less expensive. It can forego the expense, although usually modest, associated with getting a medical certificate and renewing it.

To better understand the provisions of a Sport Pilot license, it might be helpful to see how it fits with other levels of pilot licenses.

Pilot licenses include levels, ratings, endorsements, and limitations. Levels determine the "privileges", as the FAA calls them, that a pilot has. Ratings and endorsements define the kind of aircraft a pilot is authorized to fly and can add additional privileges. Limitations further define conditions that may affect the privileges of a pilot license.

The FAA issues different levels of pilot licenses. They are (from lowest level to highest level): Student Pilot, Sport Pilot, Recreational Pilot, Private Pilot, Commercial Pilot, and Airline Transport Pilot.

Each level of pilot license is said to have certain privileges. The higher the level of license, the more privileges that license allows. A pilot with a higher level of license has the privileges of a lower level of license, plus more. For example, a Commercial Pilot (speaking of someone with a Commercial Pilot license) has the privileges of a Private Pilot, plus the Commercial Pilot has the additional privilege of earning money as a pilot.

A pilot may have different levels of privileges for different kinds of aircraft. For example, a person may have Private Pilot privileges for airplanes and Sport Pilot privileges for gyroplanes.

In addition to a level of privileges, a pilot license specifies the kind of aircraft a pilot is authorized to fly. At the broadest level, those aircraft are specified by Categories and Classes.

The terms Category and Class are used both in reference to aircraft certification as well as airman certification, but there's a distinction between their meanings in those two references. For aircraft certification, Category is used with "Class" and sometimes "Purpose" to define how a grouping of aircraft may be employed, including their intended purpose or operating limitations. For airman certification, Class and Category designate the kinds of aircraft a pilot is authorized to fly.

For the purpose of airman certification, there are seven categories of aircraft and most of these categories have subordinate classes. Two examples of categories are Airplane and Rotorcraft. The Airplane category has four classes. The Rotorcraft category has two classes: Helicopter and Gyroplane.

A Sport Pilot license allows a person to fly an aircraft of the category/class for which they have demonstrated proficiency—and if that aircraft meets the definition of a light-sport aircraft.

In FAR Part 1, light-sport aircraft are defined by weight, performance, and other characteristics. Aircraft that meet the definition of light-sport aircraft in Part 1 may be flown by Sport Pilots even if the aircraft is not specifically certificated in the Light-Sport category. In terms of airman certification, however, the definition excludes certain categories/classes of aircraft from being eligible to be flown with a Sport Pilot license, reflecting a view that those kinds of aircraft are too complex to be flown with that level of license. For example, airplanes with more than one engine can't be flown with a Sport Pilot license nor helicopters or powered lift aircraft.

Below is a list of all the categories/classes of aircraft the FAA uses for pilot certification. In this list, the indented entries are classes of the category immediately above them. With the proper endorsement, a Sport Pilot may fly any of these categories/classes of aircraft except for the ones on the list that are struck through.

Categories/Classes of Aircraft
Used for Pilot Certification

Airplane - a Category with four Classes
- **Single-engine land**
- **Single-engine sea**
- ~~**Multi-engine land**~~
- ~~**Multi-engine sea**~~

Rotorcraft - a Category with two Classes
- ~~**Helicopter**~~
- **Gyroplane**

Weight Shift - a Category with two Classes
- **Weight-shift-control land**
- **Weight-shift-control sea**

Powered Parachute - a Category with two Classes
- **Powered Parachute land**
- **Powered Parachute sea**

Lighter-Than-Air - a Category with two Classes

Balloon

Airship

Gliders - a Category with no subordinate Classes

~~Powered Lift~~ - a Category with no subordinate Classes

For example, using the list above, a Sport Pilot may be authorized to fly single-engine airplanes (as long as they're light-sport), but not multi-engine airplanes. To fly either a single-engine land airplane or a single-engine sea airplane, each requires its own authorization.

There are a total of fourteen categories/classes of aircraft that are used for airman certification; that is, to specify the kinds of aircraft that pilots may fly. Of this total, a Sport Pilot may fly ten of these categories/classes. To fly a specific category/class of aircraft, a pilot must be trained in, and authorized to fly, that category/class of aircraft.

Sport Pilot privileges and limitations

For a Sport Pilot license, the FAA defines its privileges and limitations in Section 61.315 of the Federal Aviation Regulations (FAR).

Section 61.315 What are the privileges and limits of my sport pilot certificate?
(a) If you hold a sport pilot certificate you may act as pilot in command of a light-sport aircraft, except as specified in paragraph (c) of this section.
(b) You may share the operating expenses of a flight with a passenger, provided the expenses involve only fuel, oil, airport expenses, or aircraft rental fees. You must pay at least half the operating expenses of the flight.
(c) You may not act as pilot in command of a light-sport aircraft:
(1) That is carrying a passenger or property for compensation or hire.
(2) For compensation or hire.
(3) In furtherance of a business.
(4) While carrying more than one passenger.
(5) At night.
(6) In Class A airspace.
(7) In Class B, C, and D airspace, at an airport located in Class B, C, or D airspace, and to, from, through, or at an airport having an operational control tower unless you have met the requirements specified in Section 61.325.
(8) Outside the United States, unless you have prior authorization from the country in which you seek to operate. Your sport pilot certificate carries the limit "Holder does not meet ICAO requirements."
(9) To demonstrate the aircraft in flight to a prospective buyer if you are an aircraft salesperson.
(10) In a passenger-carrying airlift sponsored by a charitable organization.

(11) At an altitude of more than 10,000 feet MSL or 2,000 feet AGL, whichever is higher.
(12) When the flight or surface visibility is less than 3 statute miles.
(13) Without visual reference to the surface.
(14) If the aircraft:
(i) Has a V_H greater than 87 knots CAS, unless you have met the requirements of Section 61.327(b).
(ii) Has a V_H less than or equal to 87 knots CAS, unless you have met the requirements of Section 61.327(a) or have logged flight time as pilot in command of an airplane with a V_H less than or equal to 87 knots CAS before April 2, 2010.
(15) Contrary to any operating limitation placed on the airworthiness certificate of the aircraft being flown.
(16) Contrary to any limit on your pilot certificate or airman medical certificate, or any other limit or endorsement from an authorized instructor.
(17) Contrary to any restriction or limitation on your U.S. driver's license or any restriction or limitation imposed by judicial or administrative order when using your driver's license to satisfy a requirement of this part.
(18) While towing any object.
(19) As a pilot flight crewmember on any aircraft for which more than one pilot is required by the type certificate of the aircraft or the regulations under which the flight is conducted.

In describing the privileges and limitations of a Sport Pilot license (what a Sport Pilot may and may not do), there's a lot to chew on. Here's a simplified version of the elements that are most important for most persons.

A Sport Pilot is allowed to fly only a light-sport aircraft.

A Sport Pilot isn't allowed to carry more than one passenger.

A Sport Pilot isn't allowed to fly at night, and during the day, he can fly only under Visual Flight Rules (VFR). In simple terms, VFR means a pilot must stay out of clouds or fog or anything that prevents him from seeing very far or prevents him from seeing the ground.

A Sport Pilot can't fly too high, specifically not in Class A airspace, which begins at about three-and-a-half miles above sea level, and generally speaking, not more than 10,000 feet above sea level.

A Sport Pilot can't fly in areas associated with a large airport (Class B, C, or D airspace) or an airport with a control tower unless he has an endorsement allowing it.

There's also a catchall provision that prevents a Sport Pilot from doing anything that may be prevented by an endorsement on his license.

A Sport Pilot may use a valid driver's license in place of a medical certificate, which is required for all higher levels of pilot licenses.

That last item is worth discussing some more.

Sport Pilot medical fitness

When the Sport Pilot license was introduced, it was the first pilot license that didn't require an accompanying medical certificate.

For all levels of pilot licenses above Sport Pilot, a pilot must also have the appropriate class of airman medical certificate to exercise the privileges of that level of license. The three classes of medical certificates (from the lowest class to the highest) are third, second, and first. To exercise the privileges of a Private Pilot or Recreational Pilot license requires at least a third-class medical certificate. The privileges of a Commercial Pilot license require at least a second-class medical certificate, and the privileges of an ATP license require a first-class medical certificate.

To fly, a Sport Pilot does need some evidence of suitable medical fitness. As one form of evidence, a Sport Pilot may use an airman medical certificate, at least a third-class one. To serve as evidence of medical fitness, a Sport Pilot's medical certificate must be current.

An airman medical certificate is issued with a class (third, second, or first) and the date of the medical examination. A medical certificate is current only for a specific duration from the date of the examination. Generally, the higher the class, the shorter the certificate's duration. However, after that duration, a medical certificate may effectively revert to a lower class for an additional period of time.

In addition to its class, the duration of a medical certificate depends upon the age of its holder.

For example, consider a 30-year-old pilot who's issued a second-class medical certificate, which has a duration of 12 months. After 12 months, the certificate effectively reverts to a third-class certificate for the next 48 months. After a total of 60 months, the certificate expires.

As an alternate to a medical certificate, a Sport Pilot may use a driver's license—as long as it's valid. "Valid" means that it's the Sport Pilot's own driver's license, it's current (not expired or suspended), the address is correct, and the Sport Pilot complies with any restrictions or limitations of the license and with any administrative or judicial order the license may have that's associated with operating a motor vehicle.

Also, whether or not a Sport Pilot can use a driver's license in lieu of a medical certificate is dependent upon the state of that person's airman medical certificate, if he ever had one, or his application for a medical certificate if he ever made one. If he has or did have one, his most recent medical certificate must not have been suspended or revoked. If it was a Special Issuance medical certificate, it must not have been withdrawn. If he has applied for a medical certificate, he must have been found eligible for at least a third-class medical certificate at the time of his most recent application.

A suspended or revoked medical certificate isn't the same thing as one that has "lapsed"; that is, its expiration date has passed. A pilot who had a medical certificate and simply allowed it to lapse is eligible to use a valid driver's license with a Sport Pilot license.

Furthermore, even with a valid driver's license or a current medical certificate, a Sport Pilot must be medically fit to operate a light-sport aircraft. The FAA puts it this way: "A person ... must ...not know or have reason to know of any medical condition that would make that person unable to operate a light-sport aircraft in a safe manner."

To shorten repeating the conditions necessary for using a medical certificate or driver's license with a Sport Pilot license, I'll encapsulate those conditions with the term "medically fit".

Statement of Demonstrated Ability (SODA).

In regard to the subject of medical fitness, the case of Vance Breese is noteworthy. Vance is a gyroplane pilot with a Private Pilot license, which means that he has Sport Pilot privileges with gyroplanes, and more. To exercise the privileges of his Private Pilot license, however, requires that he have an airman medical certificate, which he does.

None of that is especially remarkable until you learn that Vance was blinded in one eye and sustained traumatic brain injuries when he was injured while attempting to set a motorcycle land speed record. Vance regained his physical abilities, and when talking with him, you wouldn't know he had a history of brain injury. Yet that blindness and history was a hurdle when Vance turned his interest to gyroplanes and sought an FAA medical certificate. It was a hurdle, but one than Vance was able to overcome. Vance tells the story of how he was able to accomplish that.

Being blind in one eye and having multiple traumatic brain injuries precluded me from getting an FAA medical in the traditional way. Instead, I pursued a process for a Statement of Demonstrated Ability (SODA) where I could be tested to determine if I had the ability to fly safely.

For two years I exchanged paperwork with the FAA surgeon assigned to my application and took multiple tests at his request. He called me one day to tell me he was going to do me a favor and reject my application.

I persisted, though, and asked that I be allowed to take a flight test to demonstrate I had the proficiency to fly safely. I argued that it would be easier for him to let me attempt the test and fail it than to have to deal with my ongoing efforts. He agreed.

For the flight test, the surgeon chose a Designated Pilot Examiner and stipulated a ninety-day time frame for the demonstrated-ability medical check ride. The surgeon created a thick booklet on all the things the DPE was to test me on, including distance measurement, multitasking, and over compliance in particular.

The day for the check ride came. I took off with the examiner and flew in an unfamiliar gyroplane to an area he designated. It had five airports, all operating with the same radio frequency. During the flight, the examiner asked me how far it was to a landmark. When I replied in feet, he asked me to count down in yards as we neared the landmark. It was a way of adding to my multitasking load while evaluating my depth perception.

At one airport, the examiner directed me, repeatedly, to land on a specific runway, but I refused. There was a gyroplane sitting at the end of that runway, spooling up its rotor, and I felt it was unsafe for me to land there with those conditions. Besides, it was against the rules. This was the examiner's way to see if I was too compliant in making aviation decisions.

I was able to pass the check ride and obtain a Statement of Demonstrated Ability. Now when it's time to renew my medical certificate, I present the SODA to an authorized aviation medical examiner, and with it my vision limitations and history of head injuries doesn't prevent me from getting a renewed medical certificate.

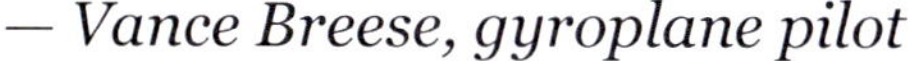

— Vance Breese, gyroplane pilot

Pilots with a higher level of license may be Sport Pilots also

Sport Pilots are allowed to fly only light-sport aircraft. Also, a Sport Pilot must be authorized to fly the particular kind of light-sport aircraft he chooses to fly. For example, a Sport Pilot must be endorsed to fly gyroplanes before flying a light-sport aircraft of that category and class.

Because a higher-level license has the privileges of a lower-level license, a pilot with a higher-level license can exercise the privileges of a Sport Pilot. As another way of putting it, he may drop down to fly as a Sport Pilot without being explicitly licensed as a Sport Pilot.

"Dropping down" to exercise the privileges of a lesser class of license applies only for the same category/class of aircraft the pilot is authorized to fly. For example, a Commercial Pilot licensed only for helicopters can't fly as a Private Pilot in airplanes.

Dropping down to fly as Sport Pilot incurs the loss of some privileges. For example, unless he has a special restriction preventing night flying, a Private Pilot may fly day or night, whereas a Sport Pilot is limited to daytime flying. As another example, a Private Pilot may, with additional training and testing, add an IFR rating to his license that allows him to fly in instrument meteorological conditions (IMC), whereas a Sport Pilot may not.

On the other hand, dropping down from a higher-level license to fly as a Sport Pilot may carry with it some privileges that otherwise require an endorsement for a pilot who has only a Sport Pilot license. For example, a pilot with only a Sport Pilot license can't fly in Class B, C, or D airspace unless he has been trained to do so and has an endorsement allowing it. However, a Private Pilot already has the privilege of flying in those airspaces and that privilege is carried down to the Sport Pilot level.

Unless otherwise indicated, I'll use the term Sport Pilot to include persons with a Sport Pilot license specifically as well as pilots with a higher-level license exercising Sport Pilot privileges.

With a higher-level license, to fly as a Sport Pilot in a particular kind of light-sport aircraft requires that the higher-level license have a rating in that kind of aircraft. For example, for someone who has a higher level of pilot license and wants to fly a light-sport airplane for land operations as a Sport Pilot, it would require that person's higher-level license to have an airplane single-engine, land (ASEL) rating. As another example, a person with a higher-level license and with a rating in gyroplanes could fly as a Sport Pilot in a light-sport gyroplane.

Unlike higher levels of pilot licenses, a Sport Pilot license doesn't include "ratings" for specific category/classes of aircraft. Instead, a Sport Pilot is issued "endorsements" permitting Sport Pilot privileges for specific category/classes of aircraft. Just as a higher level of pilot license may include more than one rating, a Sport Pilot license may include more than one endorsement.

Other endorsements may be required to fly a specific kind of light-sport aircraft. For example, a tail wheel endorsement is required to fly a tail dragger whether or not it's a light-sport aircraft.

In summary, any individual aircraft, if it fits the definition of a light-sport aircraft according to Part 1 of the FAA regulations, may be flown with a Sport Pilot license, assuming the license allows flying that category and class of aircraft and the license has any necessary endorsements that may be required for a specific aircraft.

A Sport Pilot license may be a perfect fit

In exchange for the more relaxed requirements for a Sport Pilot license, including the possibility of using a driver's license instead of a medical certificate, a Sport Pilot license has lesser privileges compared to higher classes of pilots' licenses.

For some persons, a Sport Pilot license could serve as a steppingstone to a higher-level pilot license. The Sport Pilot license typically requires less time and is less expensive to achieve that a higher-level license, and with it, a pilot can accumulate valuable experience that would be useful when pursuing a more advanced license.

For other persons, even those with a higher-level license, the privileges of a Sport Pilot license could be just right. Even with its lesser privileges and its limitations, a Sport Pilot license is ideally fitted for many pilots who fly only for leisure, including the stereotypical "weekend flyer".

With this overview of the Sport Pilot license, the next chapter discusses more specifically what it takes to become a gyroplane Sport Pilot.

Becoming a Gyroplane Sport Pilot

To fly any aircraft requires an airman pilot certificate, more commonly called a pilot license. To fly a light-sport aircraft requires either a Sport Pilot, or a higher-level, license. A Sport Pilot license requires training leading to an endorsement to fly a specific category/class of light-sport aircraft.

Sport Pilot license or another level of license?

For persons who want to fly only light-sport gyroplanes, there are usually few practical reasons for a higher-level license than Sport Pilot. Let's discuss, though, some considerations that might make a higher-level license preferable.

Fundamentally, a Sport Pilot license doesn't include the privilege of flying in certain airspaces, those designated as Class B, C, and D, and which are usually associated with controlled airports. The privilege of flying in those airspaces is included in higher-level licenses beginning with Private Pilot. However, a person with a Sport Pilot license can gain that privilege by getting additional training leading to an endorsement.

A Sport Pilot license doesn't permit flying at night. To do that, a person needs a higher-level license, at least Private Pilot. To fly at night, a gyroplane must be equipped with the necessary lights and instruments. Most light-sport aircraft aren't equipped for night flying although they can be retrofitted with the necessary equipment.

Another possible consideration for getting a higher-level license is if it's a person's ambition is to become an instructor. If a person wants to become an instructor qualified to instruct for all levels of pilot licenses applicable for gyroplanes, that person will need a Commercial Pilot license as a prerequisite leading to a Certified Flight Instructor (CFI) certificate that will allow him to instruct students for all levels of pilot licenses, including Sport Pilot, and also allow him to instruct other persons seeking to become an instructor at all levels. However, with only a Sport Pilot license, a person can obtain a flight instructor certificate that allows him to instruct to the Sport Pilot level. I'll designate that instructor certificate as CFI-Sport. A CFI-Sport certificate also allows persons to instruct other Sport Pilots aspiring to have a CFI-Sport certificate.

An important consideration in making a decision of whether or not to pursue a higher-level of license is this: the dual flight instruction time you receive from an CFI-Sport instructor can't be applied toward a higher-level license. Although some of the flight hours you accumulate may count toward the requirements for a higher-level license, the instructional time with an instructor who has a CFI-Sport certificate doesn't count toward the instructional time required for a higher-level license.

Probably the most important consideration in deciding whether or not to pursue a higher-level license is the one regarding a medical certificate. All higher levels of pilot license require a medical certificate, but a Sport Pilot license doesn't. Neither does a CFI-Sport certificate. For both the Sport Pilot license and CFI-Sport certificate, a valid driver's license may be used in place of a medical certificate, assuming you're medically fit.

For a review of what it means to be medically fit, see the chapter entitled "The Sport Pilot License".

Those are some of the considerations in making a decision of whether or not to pursue a higher-level license to fly gyroplanes. But it doesn't have to be an either-or decision. You can choose to get a Sport Pilot license and once you obtain that, pursue a higher-level license.

The first step toward becoming a gyroplane Sport Pilot

The first step toward becoming a Sport Pilot is deciding you want to become one. To help you with that decision, learn as much as you need to learn about gyroplanes and the sport of flying them. This book doesn't claim to have everything you might want to know. And it isn't intended to convince you that gyroplanes are for you. Gyroplanes aren't for everyone, just as flying isn't for everybody. But if you possibly feel an itch to fly gyroplanes, scratch it enough to learn more about them.

Where do you go to learn more about gyroplanes? A collection of leads to some information resources are listed in this book (see "Information Resources") and the companion website for the book has information about subjects related to gyroplanes. The website's address is

www.LightSportGyroplanes.com

One source for information about gyroplanes is the Popular Rotorcraft Association (PRA). Its website is www.pra.org. The organization embraces popular helicopters as well as gyroplanes, but it's predominantly about gyroplanes. Two especially valuable resources at the PRA website are its list of gyroplane instructors and information about the online ground school offered by the PRA. It's the only group ground school that I know of that's oriented toward a gyroplane perspective.

The Rotary Wing Forum at www.rotaryforum.com is an online gathering of rotorcraft enthusiasts. Although many members of the PRA participate in the Rotary Wing Forum, it isn't officially associated with the PRA. As with any open forum, be careful about opinions and advice found there. Although there's a core of knowledgeable and experienced persons who participate on the Rotary Wing Forum, there are also participants whose advice may be questionable.

The admonishment to be careful to whom you listen extends to any group. You'll likely find an abundance of homespun authorities everywhere who may, or may not, know what they're talking about.

The Rotary Wing Forum has a collection of individual forums about different topics of discussion. One of those is "Events" where you can find information about PRA chapter meetings, fly-ins, and other gatherings where you can expect to see gyroplanes and/or meet persons with an interest in them.

A firsthand way to learn more about gyroplanes and the sport is to rub elbows with gyroplane owners and pilots. Currently, nearly all gyroplane pilots are flying ones that are custom built or built from kits. Many aircraft builders belong to the Experimental Aircraft Association. EAA chapters are located across the country. Find one near you and pay them a visit. See if there's anyone there who's into gyroplanes. Even if you don't find any gyroplane enthusiasts at an EAA chapter, you can learn a lot about building aircraft if you decide that's a route you want to take.

Some of the largest events devoted primarily to gyroplanes are the PRA's gathering at Mentone, Indiana; Bensen Days at Wauchula, Florida; and the Ken Brock Freedom Fly-in at El Mirage, California. You may also find light-sport gyroplanes at larger national events such as the EAA's AirVenture held in Oshkosh, Wisconsin, and Sun n Fun in Lakeland, Florida.

Be aware that occasionally the type of flying you see at chapter meetings and even more especially at fly-ins, tend to be of the show-off variety. Don't accept this as normal flying. Some of the maneuvers you might see are performed by persons who may not realize the risk they're taking. Even with pilots who are experienced and do know what they're doing, that doesn't mean it's something for which everyone is capable.

If you're new to gyroplanes—and especially if you're an experienced pilot in other kinds of aircraft—take your time to learn how to fly gyroplanes normally.

Sometimes at fly-ins, people get the wrong idea of what is normal flying, but it's anything but normal. And because of that, it can lead to accidents. Learn what's normal flying, stick to a syllabus from an instructor, and progress according to the syllabus rather than jumping the gun.
— Desmon Butts, gyroplane CFI

In deciding how well gyroplanes may suit you, take an introductory lesson in one. A lesson, rather than a "ride", implies you go along in a gyroplane with a professional instructor.

Before flying gyroplanes, or anything else, it's important that you check your attitude toward flying. Some enthusiasts' attitudes have helped foster the image that gyroplanes are for renegades. Although that's mostly in the past, there's still an element of those persons around.

Learn about hazardous attitudes and Aeronautical Decision Making (ADM). You can start by reading the chapter on ADM in the gyroplane section of the *Rotorcraft Flying Handbook*. When reading that section, if you find yourself strongly disagreeing that the hazardous attitudes discussed there are indeed hazardous, either change your attitude or find another sport—for example, rattlesnake juggling.

Hazardous attitudes are not only dangerous for the person who has them, it's dangerous for other persons, and the sport doesn't need persons with those attitudes.

Information about the *Rotorcraft Flying Handbook* is included in the list of information resources in this book and the handbook itself is available for free online.

There's also the choice of flying an ultralight gyrocraft, which doesn't require a pilot license. As a practical matter, because of the weight limit for ultralights, this may not be an option for many persons, especially those among the weightier subjects. Also, all ultralights, including gyroplanes, are single-seaters and you can't receive dual flight instruction in one.

Don't ever, ever consider flying anything—gyroplanes or anything else—without adequate training. That also applies if you're an experienced pilot. Even if you've logged more flight time than Methuselah logged in a rocking chair, and even if you consider yourself the next best thing to Chuck Yeager, if you haven't flown gyroplanes specifically, get adequate training in them.

> *Sometimes people have been self-taught, and when they brag about it, it gives people the wrong impression.*
> *— Desmon Butts, gyroplane CFI*

To further aid your decision about becoming a gyroplane Sport Pilot, let's discuss the requirements to do that. In determining the requirements, let's consider two situations: someone who doesn't have a pilot license and someone who does.

A first-time Sport Pilot license to fly gyroplanes

If you don't have a pilot license, let's discuss what it takes to become a Sport Pilot to fly gyroplanes.

Fundamentally, to be eligible for a Sport Pilot license requires a certain minimum age. To be eligible for a Sport Pilot license to fly gyroplanes, a person must be at least seventeen years old. A person may begin training for a Sport Pilot license before his seventeenth birthday but isn't allowed to fly a gyroplane solo, which requires a Student Pilot license, until he's sixteen years old.

Also, to be eligible for a Sport Pilot license to fly gyroplanes, as is the case for all pilot licenses, a person needs to be able to speak, read, write, and understand English. If a person can't do that because of medical reasons, it may still be possible for that person to get a Sport Pilot license with some limitations placed on the license.

> **Before you begin flight training, expect an instructor to ask you for proof of US citizenship. That's usually accomplished by presenting a birth certificate accompanied by photo identification such as a driver's license or a passport.**

To earn a first-time Sport Pilot license for any kind of aircraft, a person must demonstrate an acceptable degree of knowledge and flying skills. A Sport Pilot candidate has to log at least a minimum amount of flight training in the category/class of aircraft for which he seeks authorization to fly and he must demonstrate a sufficient degree of flying skill in that kind of aircraft. Consequently, a first-time Sport Pilot license includes an endorsement permitting Sport Pilot privileges in some kind of aircraft.

To earn an initial Sport Pilot license, there are essentially three requirements:

- log at least the minimum prescribed hours of flight training,
- pass a knowledge test, and
- pass a practical test.

You must accomplish the first two requirements before you're eligible for a practical test, and training is essential for preparing you to accomplish all three requirements. Training includes ground training and flight training.

Essential ground training includes the following subjects of aeronautical knowledge:

regulations relating to sport pilot privileges, limits, and flight operations;

accident reporting requirements of the National Transportation Safety Board (NTSB);

use of the applicable portions of the aeronautical information manual and FAA advisory circulars;

use of aeronautical charts for VFR navigation using pilotage, dead reckoning, and navigation systems, as appropriate;

recognition of critical weather situations from the ground and in flight, wind shear avoidance, and the procurement and use of aeronautical weather reports and forecasts;

safe and efficient operation of aircraft, including collision avoidance, and recognition and avoidance of wake turbulence;

effects of density altitude on takeoff and climb performance;

weight and balance computations;

principles of aerodynamics, powerplants, and aircraft systems;

Aeronautical Decision Making (ADM) and risk management;

preflight actions that include how to get information on runway lengths at airports of intended use, data on takeoff and landing distances, weather reports and forecasts, and fuel requirements; and

how to plan for alternatives if the planned flight cannot be completed or if you encounter delays.

There isn't a minimum amount of time prescribed for ground training.

Incorporated with ground training, the essential flight training includes:

preflight preparation;

preflight procedures;

airport operations;

takeoffs, landings and go-arounds;

performance maneuvers;

ground reference maneuvers;

navigation;

slow flight;

emergency operations; and

post-flight procedures.

An initial Sport Pilot license with an endorsement to fly gyroplanes requires a minimum of 20 hours of flight training in light-sport gyroplanes. At least 15 of those hours must be flight training with an authorized instructor (often called "dual" instruction). At least 5 hours must be solo flight training in a light-sport gyroplane under the supervision of an authorized instructor.

Included in those 20 hours of flight training, you must log at least

10 takeoffs and landings to a full stop, and with each landing involving a flight in the traffic pattern at an airport;

2 hours of cross-country flight training;

and a solo cross-country flight that consists of

a total distance of at least 50 nautical miles,

a full-stop landing at a minimum of two points,

and during which at least one segment of the flight has a straight-line distance of at least 25 nautical miles between the takeoff and landing locations.

The minimum hours of flight training must include at least two hours of dual instruction within a certain time before taking the practical test. Those hours preparing you for the practical test must be logged within the two calendar months preceding the month you take the test. For example, if you take the practical test in July, you must have logged at least two hours of dual time since the first day of May of that year.

Your training leading to a Sport Pilot license includes both ground training and flight training with an authorized instructor. Ground training is designed to help you pass the knowledge test.

The knowledge test for a Sport Pilot license is the same no matter what kind of light-sport aircraft you intend to fly. The Sport Pilot knowledge test consists of 40 questions, all multiple-choice. The minimum passing score is 70 percent. The test, which is taken on a computer, is administered at FAA authorized testing centers. Two companies are authorized to offer knowledge tests: LaserGrade and CATS. Between those two providers, there are hundreds of locations across the country where you can take the Sport Pilot knowledge test. Before you take it, you should call to make an appointment, and you can ask the provider any questions you may have, including the cost for taking the test. You may be eligible for a discount if you belong to certain organizations; for example, the Experimental Aircraft Association or the Aircraft Owners and Pilots Association.

In case you're curious, the questions making up a knowledge test aren't always the same questions. The questions on an individual test are

selected by a computer from a larger bank of test questions and placed in random order.

Before you're allowed to take the knowledge test, you'll need an endorsement to take it from an authorized instructor. The endorsement is made by the instructor who's responsible for your ground training and that's often the same instructor who's responsible for your flight training. The written endorsement is a statement from your instructor that you're prepared to take the test. The requirement for an endorsement can save time and money in the case of some persons who may think themselves ready for the test but aren't.

After taking a knowledge test, you should get a printout of the questions you missed. Specifically, you get a printout of the Learning Codes associated with the questions you missed and these codes identify the related topics of those questions.

In preparing for a knowledge test, an authorized instructor can help you prepare for it as part of your ground school training. Often, an instructor will refer you to "home study" material where you can study on your own in preparation for the test. There are numerous sources for study aids for the Sport Pilot knowledge test, including books, DVDs, computer software and online courses. You may find scheduled, instructor-led group courses at places such as flight schools at airports or at junior colleges.

For some kinds of aircraft, especially airplanes, there are abundant study aids to help you prepare for the knowledge test as well as the oral part of the practical test during which you can expect to be asked questions about the kind of aircraft for which you're seeking an endorsement. However, study material slanted toward gyroplanes aren't nearly so abundant. The Gyroplane portion of the Rotorcraft Flying Handbook, which was developed by the FAA, is helpful. You can view it online or download it for free. You can purchase a print version from places like Amazon or order a print version from a bookstore.

The Popular Rotorcraft Association sometimes offers an online ground school specifically slanted toward gyroplanes.

Assuming you're prepared and have your instructor's endorsement, you can take the knowledge test at any time during your training: before flight training, after it, or during it. In terms of satisfying the requirement for taking a practical test, a passing grade on the knowledge test is good for two years (for up to 24 months from the month in which you passed the test), yet it's best not to wait that long. By waiting too long, some of the things you knew for certain when you took the test may escape your memory when you take the practical test and could be things the examiner may quiz you about during the practical test.

So, what happens if you should flunk the knowledge test? The answer is, you're allowed to take it again. Before you do, though, your instructor can help you better prepare, especially with the things you missed on the test. Before you take the test again, you'll need an endorsement from him that's he's helped you with the things you missed and that you're ready to retake it.

Flight training for a Sport Pilot license includes solo flying. Before you're allowed to fly solo, you'll need a Student Pilot license. Most persons get a Student Pilot license before beginning their dual flight training. A Student Pilot license leading to a Sport Pilot license, like the Sport Pilot license itself, allows you to use a valid driver's license in lieu of a medical certificate. If you choose to use a driver's license, you may apply for a Student Pilot license at an FAA Field Service District Office (FSDO) or with a Designated Pilot Examiner (DPE). A DPE is an authorized representative of the FAA, but isn't directly employed by it. You can expect to pay a fee to a DPE for the application; at the FSDO, it's free. If you choose to use a Medical Certificate, you can apply for a Student Pilot license in conjunction with applying for a medical certificate, which is done with an Airman Medical Examiner (AME).

Ground training and flight training together are intended to prepare you for a practical test. After you've passed the knowledge test and completed at least the minimum requirements of flight training, and when your instructor agrees that you're prepared, he'll endorse you to take the practical test. A practical test is administered by a Pilot Examiner from the FAA or a Designated Pilot Examiner (DPE). To shorten things, let's refer to either of those persons as an "examiner".

A practical test is often called a "check ride".

At the beginning of the practical test, the examiner will ask to see some specific documents: your identification documents, your Student Pilot license, your log book, including your instructor's endorsement for taking the practical test, and the certificate of having passed the knowledge test.

The practical test is a combination of oral testing and flight testing. The oral portion is intended for you to demonstrate your knowledge of subjects related to a Sport Pilot license and gyroplanes.

During the oral portion of the practical test, the examiner will likely ask to see the printout of any items you missed on the knowledge test. If there are any—and there probably will be—you can expect him to quiz you on those things for sure. In any case, he may ask you questions about anything that was supposed to have been covered in your training.

Most of the questions you can expect to be asked during a practical test are intended to assess your knowledge in practical terms; that is, how to relate and apply what you've learned. For example, you might be asked, "How do you recognize when you're flying behind the power curve? How do you prevent doing that? How do you recover from it if you find yourself in that situation?"

After the oral portion of the practical test, you can move on to the flight portion, and you can expect the examiner to continue to ask you questions.

The flight testing portion of the practical test is intended for you to demonstrate your knowledge and skill for becoming a Sport Pilot, including preflight preparation and ground operations. The FAA directs examiners to assess some elements of your knowledge that it considers especially important, such as Aeronautical Decision Making, runway incursion, and the use of checklists.

The flight portion of the practical test requires a gyroplane that's suitable. For a Sport Pilot license, the gyroplane must qualify as a light-sport aircraft and all of its paperwork needs to be accurate and up to date, including its registration certificate, its airworthiness certificate along with its operating limitations, and any other documents required to legally operate it. At some point before the flight portion of the practical test, the examiner will want to see those documents. He may also want to see the gyroplane's maintenance logbook, weight and balance documentation, and any operating manual.

In preparing for a practical test, you should understand specifically what you can expect to be asked to demonstrate and the standards that will be used to assess your performance. That's all defined in a helpful document called the Practical Test Standard (PTS). It's the guidance that an examiner uses to administer a practical test. For a Sport Pilot license with a gyroplane endorsement, those standards are defined in FAA-S-8081-29. It's not some secret document; the FAA publishes its Practical Test Standards for anybody to read. By searching for that designation on the internet, it should lead you to a PDF file where you'll find the inside information of what to expect on your gyroplane check ride.

A practical test administered by an FAA examiner doesn't have a fee, but one administered by a DPE does. To conduct a practical test for a gyroplane endorsement, an examiner has to be specifically authorized to do so for gyroplanes and there's a critical shortage of examiners with that authorization.

Although some persons may not confess it, I believe everyone who faces a practical test has some degree of trepidation about it. That can be a good thing. A few butterflies can help you be more alert. If you feel overly anxious, it might help you to understand that the examiner's job isn't to fail you. This is a situation where everyone involved wants you to succeed, including the examiner. His expectation is that you've got the know-how and skills to qualify for a Sport Pilot license. After all, a professional Certified Flight Instructor vouches that you do. All you have to do is show the examiner that your flight instructor is right.

Sport Pilot privileges for current pilots

Thus far, we've been discussing the process for getting a Sport Pilot license with a gyroplane endorsement for persons who don't already have a pilot license. Let's consider now another situation: a person who does have a pilot license. If that's you, what does it take for you to gain Sport Pilot privileges for gyroplanes? The answer is, it depends upon what level of license you have and the ratings and/or endorsements you have with that license.

First, if you already have a higher-class pilot license than Sport Pilot and it includes a rating for the Rotorcraft/Gyroplane category/class, then you already have Sport Pilot privileges in them, assuming you meet the provisions of being medically fit. There is one hitch, though, if your higher-level license is a Recreational license. That hitch is discussed below.

If you have a Private Pilot or higher-level license with a gyroplane rating, you have Sport Pilot privileges in a gyroplane. You also have the privilege of flying in Class B, C, and D airspaces, which requires a separate endorsement for holders of a Sport-Pilot-level-only license. If you have a Recreational Pilot license, you must satisfy the medical requirements also, and in addition, you must have a cross-country endorsement with your Recreational Pilot license that permits cross-country flights of fifty miles or more. Even if you don't intend to fly in a gyroplane that far, you still need the endorsement. Also, with a Recreational license, to fly in Class B, C, and D airspaces, you need the training and endorsement for that.

Let's consider another situation: you have a pilot license but it doesn't authorize you to fly gyroplanes. If you already have a Sport Pilot or higher-level license, you have a rating or endorsement to fly at least one category/class of aircraft. Assuming you're medically fit, you can add an endorsement to fly light-sport gyroplanes with Sport Pilot privileges by completing a proficiency check in one.

There's a difference between a proficiency check and a practical test. Although they have the same standards for demonstrated performance, there's a difference of who may administer them. A practical test, the one for a first-time Sport Pilot, is administered by either an FAA examiner or a Designated Pilot Examiner. A proficiency check, however, can be administered by a CFI, as long as it's a different CFI than the one who endorsed you for the proficiency check and as long as that other CFI is authorized to instruct in the category/class of aircraft for which you're seeking an endorsement; in this case, gyroplanes.

Staying current as a gyroplane Sport Pilot

It's often said that a pilot license is really a license to continue learning. Once you have a Sport Pilot license, you're poised for more flying adventures. Pursue those adventures safely and responsibly.

As with higher levels of pilot licenses, a Sport Pilot license doesn't have an expiration date. It's a lifetime license unless the FAA finds cause to revoke its privileges or you voluntarily surrender it. However, flying a gyroplane, like flying any other aircraft, or even driving an automobile, is a perishable skill. To remain proficient, you need to continue flying. For those who enjoy flying, that advice may sound like a doctor's prescription to take a vacation.

All pilot licenses have currency requirements. For example, to keep your Sport Pilot privileges legally active, you must complete a Biannual Flight Review (BFR) no longer than every twenty-four months. If you allow longer than the allotted time to lapse without a BFR, you can't legally exercise your pilot license privileges until you complete the review. A BFR consists of an hour of ground school and an hour of flying. The ground school can help you brush up on topics you may not have kept up with, and the flying portion can help knock off the rust of some things you may not have practiced in a while; for example, emergency procedures. Consider it another excuse to fly.

Also, other privileges require more frequent refreshing; for example, to carry a passenger.

Another way to help keep your flying skills sharp is to pursue an additional endorsement. For example, even though you may be content to fly in wide-open spaces with your Sport Pilot license, consider getting an endorsement to fly in Class B, C, and D airspaces. Besides serving as a learning opportunity, the endorsement could come in handy sometime.

Realities of flight training

When pursuing a Sport Pilot license to fly gyroplanes, there are some realities to consider. These realities aren't intended to dampen your enthusiasm but rather to inform you of some hurdles you should know about so that you don't bump into them unexpectedly.

The first reality is this: the minimum hours of flight training for a first-time Sport Pilot license are exactly that—the least amount of hours required. Most persons log more hours than the minimum in preparing for a practical test. That's true for all levels of pilot licenses.

Persons progress at different rates, not just overall, but for different parts of flight training. When it comes to so-called stick-and-rudder skills, natural ability may account to some degree for a person's rate of progress in mastering flight maneuvers, yet pilots with average natural ability or less can become the more skilled pilots. That may be because they have to try harder, and as a result, they learn to a greater depth. More often, though, a person's rate of progress is due to circumstances such as the frequency of training and how consolidated it is. In any case, getting a pilot license isn't a race. The more practice you have, the better prepared you can be for the practical test.

There's no such thing as a natural-born pilot.
— Chuck Yeager

At one time, I thought I was a natural-born pilot. But that was before I ever attempted to fly anything. My first flight lesson quickly knocked that notion from my head.

There's far more to learning to be a pilot than so-called stick-and-rudder skills. Concepts of flight, aircraft airworthiness, regulations, weather, and assessing your fitness for flying are just a few of the things important to learn in the process of becoming a pilot. Even more important is learning how to apply those things. Simply stated, learning to use good judgment is foremost.

The number of hours you may need to be adequately prepared for the practical test is impacted by the frequency and consolidation of that training. My experience with flight training is that it's better to reasonably consolidate flight training hours rather than intentionally stretching them out. For the same number of hours, flight training is generally more effective if it's condensed within a relatively short period of time. It's better to fly several hours a week for a few weeks than to fly fewer hours a week over more weeks. The reality, though, is that sometimes things beyond your control may spread out your flight training; weather for example.

It's also a reality that the shortage of gyroplane instructors may make it difficult to schedule flight training as compactly as you may want. Because there are so few gyroplane instructors, you may have to travel a long distance for training, making it impractical to travel back and forth every day or even every week. Depending upon your circumstances, you may have to stretch your training over a longer time than you'd prefer. Possibly, you could arrange with an instructor to provide you with condensed instruction and take some vacation time or otherwise set aside some time to devote yourself to training.

A few gyroplane instructors have become barnstorming nomads, traveling the country to provide instruction. You might consider inviting a roving gyroplane instructor to your area, especially if there are other persons in the area who want training and would participate.

In my opinion, for a first license, I think it's better to stick with one instructor, but you could train with more than one flight instructor if it would better optimize your schedule. I'm been under the tutelage of many instructors while pursing different ratings, taking check rides, getting checkouts in different aircraft, satisfying mandatory flight reviews, and learning to be an instructor myself. I learned from all those instructors.

> **I don't know any flight instructors who have become rich from that profession alone. To some degree, being a flight instructor is a calling. I believe that's even more so for gyroplane instructors. Unlike some airplane instructors for example, who may toil at instructing just to accumulate hours so they can land a job with the airlines, there aren't any similar prospects for gyroplane instructors.**

> *The absolutely most satisfying thing for me is when I see students meeting their goals.*
> *— Dayton Dabbs, gyroplane CFI*

Here's another reality, or at least my view of reality: the single biggest reason for the shortage of gyroplane instructors is the regulatory disparity, which among all categories and classes of light-sport aircraft, affects only gyroplanes. In addition to setting other barriers for gyroplanes, this regulatory disparity prevents light-sport gyroplanes from being readily rented for training.

All other categories and classes of aircraft that may be flown by Sport Pilots are permitted to be issued a Special Light-Sport Aircraft (SLSA) airworthiness certificate, which by regulation allows them to be rented for training purposes—but not gyroplanes.

Effectively, because of this regulatory disparity, virtually all gyroplanes in the US that can qualify as a light-sport aircraft are certificated as Experimental; that is, they're issued an Experimental airworthiness certificate. By regulation, Experimental aircraft can't be used for compensation or hire, including being rented for flight training.

To get around this restriction, so that an instructor can charge for the use of a gyroplane he provides for training, he must seek special permission from the FAA to do that. The permission, if it's approved, is in the form of a Letter of Deviation Authority (LODA).

All of the few gyroplane instructors in the US who provide training in a light-sport gyroplane must jump through the hoops to get a LODA so they can charge for the use of their gyroplane; otherwise, they couldn't afford to provide dual instruction in it.

Because an application for a LODA is a request for an exemption from established regulations, an approval for one is a laborious process involving a stack of documents. It's a lengthy process also; for example, the application for a LODA has to be made at least 60 days before its intended use.

> **Gyroplane instructors report dissimilar experience when applying for a LODA, which seems to depend at least in part on which particular FSDO handles the application and perhaps on the particular FAA official handling the application. This irregular handling of a LODA is a bump in the relatively smoother flow of processing more common regulatory applications. The requirement for a LODA probably creates as much of a hassle for the FAA staff as it is does an instructor.**

Getting a LODA was a bit of a hassle. It took about six months. For someone like me, sitting on an expensive aircraft, that's a long time to go without income.
— Dayton Dabbs, gyroplane CFI

Another concern about LODAs is that it's issued for only a two-year period—at most. It can be yanked at any time. That's hardly a secure foundation for someone considering becoming a gyroplane instructor, especially for someone considering investing in a gyroplane to provide training, a factor that inhibits potential gyroplane CFIs.

For all other category/classes of aircraft that Sport Pilots may fly, the FAA allows those aircraft to be factory-built and certificated so that they can be used for training and the regulations allow the owner to be compensated for that use—but not for gyroplanes. Because this disparity for gyroplanes does currently exist, we have to deal with this reality as long as it's in effect.

An alternative to taking training in an instructor's two-place gyroplane is to have your own two-place gyroplane in which you can train. There's no regulation preventing you from receiving flight training from an authorized instructor in your own aircraft, even if it's an Experimental one, and the instructor can be paid for that instruction. The reality of that, though, is that you have to own the aircraft and that involves a lump of up-front financial outlay before beginning your flight training. Financially, that raises the entry barrier to gyroplane training, stopping some would-be pilots before they even begin.

Also, some instructors are reluctant to provide flight training in a student's gyroplane. One reason for that reluctance is the instructor's concern about the safety and stability of a student's gyroplane. An instructor is usually less familiar with a student's gyroplane than he is with his own. Furthermore, an instructor's own gyroplane that he uses for training, one allowed by a LODA, is required to be inspected, not just annually, but also every 100 hours, the same as any other aircraft used for training. There's no requirement, however, for a 100-hour inspection for a gyroplane owned by a student. It's not unreasonable that an instructor should be less confident in the safety of an aircraft that's not his own, especially if it's not subject to more frequent mandatory inspection.

Another reality in regard to training due to the regulatory disparity with gyroplanes is the concern with the requirement for the solo flight training necessary to get a Sport Pilot license.

As an instructor, I'm confronted with how do I get my students to solo unless they buy their own gyroplane.
— Dayton Dabbs, gyroplane CFI

A LODA only permits an instructor to collect a fee for the use of his gyroplane when giving dual flight instruction. It doesn't permit charging a student for the use of an instructor's gyroplane for solo flight training. Thus, a flight instructor incurs a direct financial cost, as well as assuming an additional and substantial financial risk, by lending his gyroplane to be flown solo by a relatively inexperienced person.

Some gyroplane instructors have adopted the philosophy that if they have enough confidence to endorse a student pilot for solo, they also should have enough confidence in that person's ability to allow the student to fly solo in the instructor's gyroplane, but not instructors share that philosophy.

I will let a student fly [solo] in my aircraft if they're ready to. Experimental aircraft aren't insurable and I assume all the risk when I do that.
— Gyroplane CFI Desmon Butts

Other than gyroplanes, there's no other category/class of aircraft that effectively requires either the instructor or student to own an aircraft in order to receive flight training.

Although there are unique hurdles to becoming a gyroplane Sport Pilot, persons who have accomplished that have found the reward worth the effort.

You can find more information about becoming a gyroplane Sport Pilot and resources to help you do that at the Light-Sport Gyroplanes website.

www.LightSportGyroplanes.com

Photo 7-1 A Calidus gyroplane above the clouds

The Future for Light-Sport Gyroplanes

In 2004, when the FAA implemented the Light-Sport Aircraft and Sport Pilot (LSA/SP) regulations, those regulations allowed Sport Pilots to fly a variety of light-sport aircraft: airplanes, gliders, lighter-than-air balloons and airships, powered parachutes, weight-shift-control aircraft, and gyroplanes. All of those categories and classes of light-sport aircraft—except for one class—were allowed to be factory-built in accordance with an industry consensus standard and certificated with a Special airworthiness certificate in the Light-Sport category (SLSA). The lone exception was the gyroplane class. Unlike all other light-sport aircraft, FAA regulations restrict gyroplanes from being factory-built and certificated as SLSA.

The year 2014 marks the ten-year anniversary of the LSA/SP regulations. Light-sport aircraft, specifically intended for Sport Pilots, have proven popular. There are now thousands of licensed Sport Pilots. Many of these persons wouldn't be flying without those regulations. The sales and services provisioning Sport Pilots have benefited the US economy.

Looking to the US, other countries have implemented similar regulations and have seen similar success. Yet, despite pioneering the Light-Sport Aircraft/Sport Pilot era, the US lags those countries when it comes to light-sport gyroplanes. That shortcoming is due to a regulatory disparity in the LSA/SP regulations, which was put there at the beginning of those regulations, and ten years later, is still stalling the full benefit of those regulations in the US. That disparity, from a practical viewpoint, prevents Sport Pilots from flying factory-built gyroplanes.

The FAA allows pilots to fly factory-built gyroplanes that are type certificated. However, the last time a type certificated gyroplane was manufactured in the US was nearly fifty years ago. Today, there are less than a dozen of these gyroplanes still in existence in the entire country, some only in museums, and few of them of them are still flyable. In any case, and more directly related to the issue, existing type certificated gyroplanes don't meet the definition of a light-sport gyroplane and therefore can't be flown by Sport Pilots.

Although an alternative for Sport Pilots who want to fly factory-built gyroplanes is to fly ones that meet the definition of light-sport in FAA Part 1 and that are certificated as Experimental Exhibition, an Experimental Exhibition airworthiness certificate imposes such severe restrictions on the operation of an aircraft—such as where and when the aircraft may be flown without advance notice to the FAA and limiting the circumstances for carrying a passenger—that it's impractical for ordinary recreational flying.

Consequently, the effective result of the confluence of existing regulations is an odd incongruity. The regulations allow some pilots, those with a Private Pilot or higher-level license, to fly Experimental gyroplanes, and they may do so without a rating in gyroplanes or even any training in them. Furthermore, the regulations allow persons who have no pilot license at all to fly factory-built ultralight gyrocraft, which aren't required to have any kind of aircraft certification. Yet, the regulations effectively prevent Sport Pilots from flying factory-built gyroplanes, even though those gyroplanes would be required to be not only certificated but would also require the pilots to have specific training in gyroplancs to fly them.

In practical terms, Sport Pilots who want to fly gyroplanes have only one recourse: gyroplanes certificated as Experimental Amateur-Built. And because Experimental aircraft aren't allowed by regulations to be rented, a Sport Pilot is further restricted in his options for flying gyroplanes. Without the option to rent a gyroplane, a Sport Pilot must make the financial commitment to purchase one in order to fly these aircraft; in other words, he must buy one to fly one.

The Case for SLSA gyroplanes

The biggest barrier inhibiting light-sport gyroplanes in the US is the regulatory disparity that restricts light-sport gyroplanes from being factory-built and certificated as SLSA. Gyroplanes are the only category/class of aircraft among all the ones that Sport Pilots may choose to fly for which this is the case. Removing this regulatory disparity will do more to grow light-sport gyroplanes in the US than anything else.

This regulatory disparity discriminating against gyroplanes not only inhibits the growth of light-sport gyroplanes in the US, it's an obstacle to improving gyroplane training and it's detrimental to gyroplane safety.

The regulatory disparity appears in Section 21.190(a) of the Federal Aviation Regulations (FAR). It reads

> **(a) *Purpose*. The FAA issues a special airworthiness certificate in the light-sport category to operate a light-sport aircraft, other than a gyroplane.**

It's those last four words in 21.190(a) that single out gyroplanes from all other categories/classes of aircraft that Sport Pilots may fly. They restrict gyroplanes from being certificated as SLSA and effectively prevent Sport Pilots from flying factory-built light-sport gyroplanes.

The SLSA airworthiness certificate

Outside of ultralights, which the FAA refers to as "vehicles", an airworthiness certificate is an essential document that's necessary for an aircraft to be legally eligible to be flown. Each individual aircraft must have its own airworthiness certificate. There are different kinds of airworthiness certificates, each kind reflecting the purpose of the aircraft. Each kind of airworthiness certificate has regulatory provisions associated with it that dictate, among other things, the provisions for which the aircraft can be used and under what provisions it can be built.

Airworthiness certificates are categorized by a hierarchy of class, category, and purpose. At the highest level, there are two classes of airworthiness certificates: Standard and Special.

Standard airworthiness certificates are issued to aircraft that are type certificated, those that can be used for the most extensive range of commercial purposes. For example, standard airworthiness certificates are issued to type certificated aircraft ranging from DC-10 airliners to Cessna 172s. The Standard airworthiness certificate class has several categories; for example, Normal, Utility, Commuter, and Transport.

The other class of airworthiness certificate is *Special*, which restricts more severely the commercial uses of aircraft. A Special airworthiness certificate may be issued for some type certificated aircraft to limit them to specific commercial applications; for example, Agricultural. More commonly, though, a Special Airworthiness certificate is issued to aircraft that aren't type certificated.

The Special class of airworthiness certificate has its own categories; for example, Primary, Restricted, Limited, Experimental, and Light-Sport. The Light-Sport category was added in conjunction with the LSA/SP regulations.

Section 21.190 of the FAR authorizes the issuance of Special airworthiness certificates in the Light-Sport category; that is, for light-sport aircraft that are factory-built in accordance with industry consensus standards.

A Special airworthiness certificate issued to an aircraft in the Light-Sport Category is termed an SLSA airworthiness certificate. By extension, those aircraft with this class and category of airworthiness certificate are called SLSA.

What's so special about a Special Light-Sport airworthiness certificate and why is it important that the FAA allow SLSA gyroplanes in the US?

The full answer to that question could require a separate book. To simplify the answer, let's look at three of the most important benefits to pilots for having SLSA gyroplanes: safer gyroplanes, improved gyroplane training, and a greater availability of gyroplanes. From a wider perspective, there's also an economic benefit with SLSA gyroplanes.

Safer gyroplanes

An SLSA airworthiness certificate is issued for a factory-built light-sport aircraft, but not simply because it's built in a factory. The aircraft must be manufactured in accordance with an industry consensus standard, a standard developed by a consensus agreement of participants from the aviation industry who collectively define the criteria that a factory-built light-sport aircraft must meet.

A consensus standard for a light-sport aircraft is developed by representatives from the aviation industry under a strict procedural regimen. In the US, that regimen is prescribed by ASTM, an internationally prominent organization that provides guidance for developing industry standards across a wide variety of industries. An overall ASTM industry consensus standard for an aircraft is actually a body of individual consensus standards that collectively define the entire spectrum of designing, producing, and sustaining an aircraft.

To ensure conformance with an industry consensus standard, a manufacturer of SLSA must meet the requirements of three essential elements of that standard: design and performance, production, and on-going support, including safety-of-flight issues.

To meet the standard for design and performance, a manufacturer must first develop a prototype aircraft, and through flight testing and other verifications, ensure that the prototype conforms to the consensus standard for that category/class of aircraft, including the aircraft's flight characteristics and stability.

Once a prototype's design and performance is confirmed to be compliant with the standard for its category/class of aircraft and the aircraft moves to production, the manufacturer must attest that each of the aircraft that comes off the assembly line is a precise replicate of that prototype. Once built, these manufactured replicates must be ground and flight tested to confirm that their performance is consistent with the consensus standard and that they're safe for operation.

Furthermore, the manufacturer must ensure its production of these aircraft is compliant with an industry consensus standard for manufacturing, including quality standards for materials, fabrication, and assembly.

After the aircraft are built and placed in operation, the manufacturer must adhere to an industry consensus standard for monitoring safety-of-flight issues, including making provision for any such issues that arise and providing a way of alerting owners of those issues.

An important benefit of having SLSA gyroplanes—ones that are factory-built and that are designed, produced, and monitored in conformance with an industry consensus standard—is that they would be built consistently to quality standards and that they would have demonstrated conformance to an industry-accepted standard of performance, leading to a greater assurance that these gyroplanes are stable and safe.

As with other light-sport aircraft, before there can be factory-built SLSA gyroplanes in the US, there must be an industry consensus standard for them. That standard for gyroplanes exists. It was developed shortly after the LSA/SP regulations were implemented and the FAA has accepted it. Yet, the disparity in FAR 21.190(a) still exists and the FAA still prevents pilots from flying factory-built, SLSA gyroplanes

It's ironic that when the LSA/SP regulations were implemented, the FAA blocked factory-built SLSA gyroplanes due to its concern that there were too many accidents involving gyroplanes, virtually all of which were amateur-built, yet by blocking factory-built SLSA gyroplanes, the result is that today virtually the only gyroplanes being flown are still amateur-built.

In 2012, the National Transportation Safety Board (NTSB) released findings from a study entitled "The Safety of Experimental Amateur-Built Aircraft". In summary, the study found that Experimental Amateur-Built (EAB) aircraft represented nearly 10 percent of the general aviation fleet, yet accounted for approximately 15 percent of the accidents. Further, the study found that EAB aircraft accounted for 21 percent of the fatal accidents during the year 2011. The NTSB report included this observation

> **E-AB aircraft vary considerably with respect to structural and performance characteristics, some of which require specific training ...**

EAB aircraft are typically built by persons who aren't professional aircraft builders. With many homebuilders, even an EAB aircraft built from a kit can vary considerably in structural characteristics due to varying degrees of builders' construction skills as well as the different choices that an individual builder may make in substituting different materials or in modifying a kit. In general, EAB aircraft are less likely to be consistently built than aircraft produced in a factory where there is a quality assurance plan for materials and with periodic inspections to ensure conformance to a defined assembly and production plan.

Also, EAB aircraft vary considerably with respect to performance characteristics. Experimental Amateur-Built aircraft are intended by regulation only for "person(s) who undertook the construction project solely for their own education or recreation". Part of that education can be constructing an aircraft to try new and unproven concepts, resulting in varying performance characteristics among EAB aircraft. Building an aircraft for the first time can introduce an additional element of danger when it comes time to fly that aircraft. To some degree, every Experimental Amateur Built aircraft can be unique and every first-time flight of a newly built EAB is a test flight.

It isn't my intention to diminish the value of the experience that amateurs gain from building experimental aircraft. I respect the notion of building aircraft for education and especially the notion of conceiving and testing new ideas. It's a part of America's renown for ingenuity.

I know persons who have built their own aircraft, either from a kit or based upon their own design, whose workmanship in construction is of a high quality. Especially with kits for which there are builders' assistance programs, workmanship is of a high quality consistently. My point in elaborating the differences between EAB and factory-built aircraft is to distinguish the lesser consistency typical of the many other EAB aircraft.

Improved gyroplane training

A second important benefit of having SLSA gyroplanes would be improved training. To understand this issue better, let's turn to Part 91 of the Federal Aviation Regulations.

FAR Section 91.327 defines the few commercial activities allowed for an SLSA. One of those activities is flight training.

Sec. 91.327 Aircraft having a special airworthiness certificate in the light-sport category: Operating limitations.

(a) No person may operate an aircraft that has a special airworthiness certificate in the light-sport category for compensation or hire except—

(2) To conduct flight training.

Although Section 91.327 refers to operating "limitations", paragraph (a) could be termed "permissions". It identifies activities for which SLSA are permitted to be used for compensation or hire; that is, activities for which owners may receive money for the use of their aircraft, one of which is for conducting flight training.

Subparagraph (a) (2) of Section 91.327 is important not only because it permits SLSA owners to be compensated for the use of their aircraft to conduct flight training, but also because the permission is explicit in the regulation. It doesn't require jumping through any other regulatory hoops to have it.

Experimental aircraft; that is, those certificated in the Experimental category, aren't allowed by regulations to be used for compensation when employed for training. It's only by exception from the regulations that they may be permitted to do so. The owner of an Experimental aircraft who wants to be paid for the use of his aircraft for training must seek permission from the FAA to allow him an exception from the regulations. Exceptions, if granted, are extended to an individual owner on a case-by-case basis, not to a whole category of aircraft.

For other kinds of light-sport aircraft—an airplane for example—an instructor can, by regulation, provide a consensus compliant, factory-built aircraft for training and be compensated for its use. The FAA allows those aircraft to be certificated as SLSA. But for gyroplanes, that's not the case. Currently, because of the disparity with regard to light-sport gyroplanes, the only gyroplane an instructor may provide for Sport Pilot training, and be allowed compensation for its use, is a gyroplane certificated as Experimental, and then only if the FAA allows the compensation for that instructor and that particular gyroplane; that is, the FAA allows it only on a case-by-case basis.

Let's look again at this observation from the NTSB report

> **E-AB aircraft vary considerably with respect to structural and performance characteristics, some of which require specific training ...**

In regard to the last phrase in the NTSB observation that some EAB aircraft require a pilot to have specific training, the NTSB study had more to say about that.

> **Both the accident analysis and extensive discussions with EAA members, kit manufacturers, and E-AB aircraft builders emphasized the importance of the builder receiving appropriate and sufficient training to develop proficiency with the new type of aircraft prior to flying his/her E-AB aircraft.**
>
> **These discussions identified challenges in finding training aircraft and instructors. Their scarcity, in part, is a result of the difficulty in obtaining an exception to the FAA regulation prohibiting a qualified instructor who owns an E-AB aircraft from charging students for instruction in the aircraft.**

These statements from the NTSB study have additional importance for gyroplanes because currently all gyroplane Sport Pilot flight training is conducted in Experimental Amateur-Built aircraft, including fundamental flight training for first-time pilots. This means that, currently, even fundamental training in a gyroplane may only equip a student to adequately fly the specific gyroplane that he's trained in and only partially prepare him to fly any other gyroplane.

Permitting SLSA gyroplanes would improve light-sport gyroplane training in three important ways. First, it would establish a "norm" for gyroplane training. Because all SLSA gyroplanes must conform to the same consensus standard for design and performance, a student pilot undergoing fundamental flight instruction would learn skills that could be consistently applied among all consensus compliant gyroplanes. Training in an Experimental Amateur-Built gyroplane—appropriate for transition training, but not fundamental training—could still be allowed for pilots who are already trained and licensed so that they could learn the unique characteristics of a particular EAB aircraft.

Second, SLSA aircraft are allowed by regulation to be rented for training. It wouldn't be necessary for an instructor to seek an exception to the regulations to be allowed to charge for the use of his gyroplane for training, one of the impediments that's a cause for the current dearth of gyroplane instructors.

A third way that SLSA gyroplanes would improve training is closely related to the second. Because SLSA aircraft are allowed to be rented for flight training, neither an instructor nor a student would have to own a gyroplane for training.

Currently, in order for an instructor to charge a student for the Experimental aircraft the instructor provides for training, the instructor must get an exception to the regulations from the FAA, usually in the form of a Letter of Deviation Authority (LODA). Although a LODA permits an instructor to charge for the use of his gyroplane when providing dual flight instruction, a LODA doesn't allow renting that aircraft to the student for solo flights. Some instructors do allow a student of their own to fly solo in the instructor's gyroplane, but many don't. It's an expense for which the instructor can't legally be compensated, plus it's a liability risk for the instructor. It's often difficult for an instructor to find insurance for instructing in an Experimental aircraft, let alone affordable insurance; consequently, not all instructors have insurance and must bear the cost of any mishap entirely on their own.

Conceptually, a student could buy a two-place gyroplane and arrange for an instructor to provide dual flight training in the student's gyroplane, and when approved by the instructor, the student could fly solo in the aircraft he owns. Since that situation doesn't involve the instructor being compensated for the use of an aircraft, but only for his instructional service, that's within the regulations. In reality, though, there are few instructors who are willing to instruct in a student's gyroplane, one with which they're less familiar than their own and less sure of its maintenance.

With current regulations in the US, for a Student Pilot to complete flight training in a gyroplane, the instructor, the student, or both must own a gyroplane. In most cases, to complete gyroplane training leading to an initial pilot license, both the instructor and the student must own a gyroplane: the instructor to conduct dual flight training and the student for solo flight training.

With SLSA gyroplanes, neither an instructor nor a student pilot would have to own a gyroplane for training. With SLSA gyroplanes, flight schools could provide those gyroplanes, as is the case typical with training in other kinds of aircraft. With SLSA gyroplanes and flight schools that own them, instructors could be hired to offer flight training in those gyroplanes and students could rent them for solo training.

With SLSA gyroplanes, which can be rented for flight training, the result would be a greater number of gyroplane flight instructors and more accessible training for persons wanting to learn to fly gyroplanes.

Greater availability of gyroplanes

The regulatory disparity in FAR 21.190(a) that prevents Sport Pilots from flying factory-built gyroplanes also effectively prevents them from renting gyroplanes.

Few student pilots own the aircraft they train in, including for solo training. Many other pilots don't own the aircraft they fly; they choose to rent one instead. Gyroplane Sport Pilots, however, don't have that choice.

With the availability of SLSA gyroplanes, which are allowed to be rented, a Sport Pilot wouldn't have to buy a gyroplane to fly one. Consequently, SLSA gyroplanes would greatly reduce the cost barrier to flying gyroplanes, especially for persons just entering the sport, making gyroplanes more widely available to a larger number of persons.

A greater availability of light-sport gyroplanes would grow the sport, but of greater importance, that growth would consist largely of a greater number of factory-built gyroplanes, ones engineered and manufactured in accordance with industry consensus standards and with consistently demonstrated stability—in short, safer gyroplanes.

Economic benefit of SLSA gyroplanes

As a result of the LSA/SP regulations, light-sport aircraft have proliferated. There are over a hundred and fifty different models of Light-Sport-certificated aircraft available in the US and thousands of Sport Pilots. The Light-Sport segment has been a relative bright spot in an otherwise dim general aviation industry. The manufacturing, employment and maintenance services associated with factory-built LSA contribute to the US economy. Also, the sales, rental, and services associated with the pilots who fly Light-Sport certificated aircraft are a boon to the economy.

Other countries with regulations similar to those of the US have seen similar economic success, yet the US falls short of those countries when it comes to light-sport gyroplanes. In other parts of the world, factory-built gyroplanes are becoming increasingly popular, a growth that's reflected not only by the increasing number of aircraft and the number of persons attracted to these unique aircraft but also by the economic impact of this still small, but emerging, segment of the aviation industry within those countries. In those countries, this growth is fostered by regulations favorable to that growth. The US, however, is missing a greater opportunity to participate in this segment of aviation and its economic benefit.

Factory-built gyroplanes are thriving in other countries. To allow US Sport Pilots to have access to these gyroplanes, some of those manufacturers offer kits in the US, even to the extent of re-engineering factory-built aircraft so they can be kit-built. A larger number of manufacturers, however, are holding off entering the US market because of the regulatory disparity preventing factory-built, certificated SLSA gyroplanes in the US. Allowing SLSA gyroplanes would be an economic benefit to general aviation as well as the greater US economy.

SLSA: the future for light-sport gyroplanes

The year 2014 marks the ten-year anniversary of the LSA/SP regulations. The regulatory disparity that discriminates against light-sport gyroplanes was put in place at the beginning of those regulations and, ten years later, is still stalling the full benefits of these aircraft to US Sport Pilots.

In addition to its detrimental impact on safety, training and growth, the disparity in the FAA regulations that prevents factory-built SLSA gyroplanes also stymies the economic contribution of these aircraft.

It's past time that the FAA enable these benefits and allow for gyroplane pilots the same opportunity it allows for pilots of all other categories and classes of light-sport aircraft: to fly gyroplanes that are factory-built to industry consensus standards. The FAA can extend that freedom to gyroplane pilots by removing the current regulatory disparity that prevents SLSA gyroplanes.

At the following website, you can find a fuller explanation of this disparity and its remedy. There also, you can follow other issues, developments, and events as the future for gyroplanes unfolds.

www.LightSportGyroplaneS.com

Information Resources

Books

Note: There are few books about gyroplanes, especially current ones. Some of the books in the following list are out of print, but may be found with vigorous internet searching. Also, when found, a few of these books sell at prices associated with rare books.

Abbott, Paul Bergen: *The Gyroplane Flight Manual*, Indianapolis, The Abbott Company, 1997—1996.

Abbott, Paul Bergen: *Understanding the Gryoplane: A Handbook of Technical Information for the Non-Technically-Minded*, Indianapolis, The Abbott Company, 1988, 1994.

Bensen, Igor B.: *A* Dream *of Flight*, Indianapolis, The Abbott Company, 1992, 2003.

Charnov, Bruce H.: *From Autogiro to Gyroplane: The Amazing Survival of an Aviation Technology*, Westport, Connecticut: Praeger Publishers, 2003.

Chiles: James R.: The *God Machine*, New York, Bantam Dell, 2007.

FAA: *Pilots Handbook of Aeronautical Knowledge* (FAA AC 61-23C).

FAA: *The Rotorcraft Flying Handbook* (FAA-8083-21), US Dept. of Transportation, Federal Aviation Administration, 2000. (available online as a PDF file)

Harwood, Philip A.: *Flying a 'New Generation' Gyrocopter: A guide for converting pilots*, The Gyrocopter Company UK Ltd, 2008.

Hollman, Martin: *Flying the Gyroplane*, Monterey, California, Aircraft Designs, Inc., 1984.

Hollmann, Martin: *Modern Gyroplane Design*, Monterey, California, Aircraft Designs, Inc., 1992.

Howell, Patrick: *Gyroplane Flying for Beginners*, Alcester, Warwickshire, UK. (on CD-ROM).

O'Connor, Timothy S.: *You Can Afford to Be a Pilot*, 2010.

Organ, Dave: *An Introduction to Ultralight Gyroplanes*, Apex Publishing, Cheltenham, UK, 2002.

Patterson, Lee and Patterson, Bob: *The Gyroplane In Law Enforcement*, Kindle edition, 2012.

Springer, Marion: *Born Free: My Life in Gyrocopters*, published by Springer Enterprises, 2004.

Periodicals

Kitplanes magazine, Box 8535, Big Sandy, Texas 75755.
Website is www. Kitplanes.com

Powered Sport Flying magazine, Sport Aviation Press LLC, Box 38, Greenville, Illinois 62249. (includes a Rotorcraft section) Monthly. Website is www. PSFmagazine.com

World Directory of Leisure Aviation, Will Tacke. Published annually. Distributed in the US in connection with *Powered Sport Flying* magazine.
Website is www.psfmagazine.com/world-directory-of-leisure-aviation/

Organizations

Experimental Aircraft Association (EAA), EAA Aviation Center, 3000 Poberezny Rd., Oshkosh, Wisconnsin 54902, (800) 564) 6322. Website is www.EAA.org

Light Aircraft Manufacturers Association (LAMA), 2001 Steamboat Ridge Ct., Daytona Beach, Florida 32128, (651) 592-7565. Website is www.lama.bz

Popular Rotorcraft Association, (574) 353-7227. Website is www.PRA.org
The PRA website has a list of local chapters.

Online resources

Federal Aviation Administration: www.FAA.gov

FAA website for finding Field Service Districts (FSDOs)
www.faa gov/about/office_org/field_offices/fsdo/

Sport Pilot Practical Test Standards: (a PDF file from the FAA)
www.faa.gov/training_testing/testing/test_standards/media/faa-s-8081-29.pdf

National Transportation Safety Board (NTSB): *The Safety of Experimental Amateur-Built Aircraft*. Available at www.ntsb.gov/doclib/reports/2012/SS1201.pdf

Sport Pilot (EAA): www.SportPilot.org

Other resources

See the Gyroplane Directory that follows for information and data about gyroplanes available in the US and for contact information for manufacturers and agents.

www.LightSportGyroplanes.com

Gyroplane Directory

The Gyroplane Directory attempts to provide, as nearly as feasible, a direct comparison of different models of gyroplanes available in the US by listing the same data items for each model in the same order

The data for each model was provided directly by the gyroplane manufacturer or its agent; however, the currency and accuracy of that data as it appears in the Directory is not guaranteed. For the most current information and accurate data about any model listed in the Directory, contact the manufacturer or its agent directly.

The models listed in the Directory are representative of gyroplanes available in the US and not all manufacturers or models are listed. For those manufacturers that are included, not all of them provided complete data for all models. In some cases, a manufacturer provided no data directly, and in those cases, no data is listed.

Unless noted, the models listed in the Directory are ones that are available as complete kits. Models that are available only as partial kits or as plans-only are not included.

Readers' guide to the Gyroplane Directory

The Gyroplane Directory is arranged in alphabetical order by the names of the manufacturers. For each manufacturer, the first page contains contact information for that manufacturer and a brief summary of the models listed in the Directory. Following the page with the contact information is the data for each of those models. In the case where a manufacturer provided no data, only the page with the contact information appears.

The directory information includes information to help determine if the model meets the requirements to qualify as a light-sport aircraft in the US. For a gyroplane to qualify as an LSA, it must meet the requirements defined in Part 1 of the FAA's Federal Aviation Regulations (FAR) that are applicable to gyroplanes. (For a discussion of those requirements, see the chapter "Light-Sport Aircraft".)

The following further defines the data items listed in the directory.

Models The name or designation of the model that heads each column of data about that model.

Light-Sport (Yes/No) "Yes" or "No" indicates whether or not the model, as specified by the manufacturer, satisfies the requirements for a light-sport aircraft; that is, its eligibility to be flown with a Sport Pilot license.

First year of production This is the calendar year in which the model began production. This may be different than the year in which the model was introduced, perhaps as a mockup, demonstration model, or production prototype. If the model is not yet in production, an anticipated date of production is given, if known.

Number produced This is the number of this model that have been completed, either as factory-built or as a kit.

Max number of occupants This is the maximum number of occupants for which the model is designed, including the pilot.

Seating arrangement Seating arrangement is defined as
"Single" (pilot only),
"Tandem" (two seats with one seat in front of the other, and unless otherwise noted, the pilot's station is in the front seat),
or "Side-by-side" (two seats aside each other).
Any other seating arrangement is described specifically.

Cockpit enclosure A cockpit enclosure is defined as "None" if it has no overhead covering, pod, side panels or other enclosure.
"Open" refers to a cockpit with no overhead covering but may have a pod that encloses the front and/or sides.
"Enclosed" is defined as fully enclosed; e.g., with two side doors or an overhead canopy.
"Convertible" may be noted also if some enclosures may be optionally removed.
If the enclosure is other that the above, it is specified.

Engine(s) The engine provided as standard with the model appears first, followed by any optional engine(s) offered directly by the manufacturer. Engine(s) refers to the engine powering the propeller. In rare instances, such as a fly/drive model, it may have another engine used to power the wheels when functioning as a ground vehicle.

Starter This is the type of starter: electric or pull. Some models with smaller two-stroke engines may not have electric starters.

Note about rotor blades: unless otherwise noted, each model comes equipped with a fixed-pitch, semi-rigid, teetering, two-blade rotor system.

Rotor blade manufacturer The name of the rotor blade manufacture is given.

Rotor blade material The material of which the rotor blades are made.

Rotor disk diameter The rotor disk diameter, given in feet (unless otherwise noted), is the span of the two rotor blades, including the hub.

Rotor blade cord The rotor blade cord (width) of the blades is given in inches.

Prerotator If equipped with a prerotator, the kind is given; e.g., mechanical, hydraulic, electric.

Rotor brake If a rotor brake is included, this is noted with "Yes".

Propeller Information provided for the model's propeller includes
how the propeller is mounted: as a pusher (pointed toward the rear) or as a tractor (pointed toward the front).
the number of blades,
the propeller's diameter, given in inches,
the material of which the blades are composed (e.g., wood, metal, composite),
information about a propeller reduction drive, including its kind (e.g., belt or gear), and its reduction ratio is given as the ratio of engine revolutions to one propeller revolution.
Other information includes whether or not the propeller is ground adjustable, and if ground adjustable, the range of pitch adjustment.
Note: In some cases, an in-flight adjustable propeller may be offered as an option.
Lastly, the propeller's manufacturer is given.

Airframe This is a description of the airframe's material and construction; e.g., welded or bolted.

Empty weight Empty weight is given in pounds (lb). Empty weight is the basic weight of the model as it comes from the manufacturer, including installed equipment, plus the weight of fluids such as oil and gas that are not usable.

Maximum Takeoff Weight (MTOW) MTOW is given in pounds.

Useful load Useful load is given in pounds. Useful load is the gyroplane's MTOW minus its Empty Weight. Useful load is the most load the gyroplane can carry, including the pilot, other occupants, baggage, and fuel. Note: When considering useful load, it is also important to consider the gyroplane's entire weight and balance. It may be possible to load a gyroplane with less than its useful load yet, because of where the weight is placed, make the gyroplane dangerously unbalanced.

Maximum fuel capacity The most fuel the model's fuel tank(s) are designed to hold (in gallons). Some models offer optional, auxiliary fuel tanks. If so, that option is noted and the quantity of fuel the auxiliary tanks can hold is given.

Unusable fuel Not all of the fuel a tank can hold (its maximum capacity) can be used. The amount of unusable fuel is given in gallons.

Fuel The fuel required for the model's engine or the recommended fuel for that engine. In some cases, the manufacturer may specify an alternative fuel, perhaps recommended only if the primary fuel is not available. A common example is a gyroplane with a Rotax 912 or 914 engine where the primary fuel recommended by the gyroplane's manufacturer may be automotive fuel, but the manufacturer specifies Avgas 100LL may be used if automotive gas isn't available. Note: Manufacturers offering the same engines may specify different fuel requirements for their models. Using a fuel not specified by the manufacturer may be a warranty concern.

Maximum speed (V_{NE}) V_{NE} is the maximum speed specified for the model and should never be exceeded. V_{NE} is given in knots.

Maximum continuous cruise The maximum airspeed in level flight with maximum continuous power (V_H). V_H is given in knots.

Maximum range Range is given in statue miles. This is the maximum distance the gyroplane can fly, and unless otherwise noted, with no useable fuel remaining in the tanks; that is, without any fuel reserve. The power setting for Max Range is given as a percentage.

Maxlmum endurance Maximum endurance is given in hours and minutes. Unless otherwise noted, this is the most time the gyroplane can fly without any fuel reserve. Note: FAA regulations require a fuel reserve for cross country flights.

Takeoff distance Takeoff distance is given in feet. It is the distance from the beginning of a takeoff roll to wheels off under standard conditions.

Maximum rate of climb This is the maximum rate of climb (expressed as feet per minute) at maximum takeoff weight and with standard conditions.

Landing distance This is the minimum ground roll for a landing in standard conditions. Landing distance is in feet.

Length Unless otherwise noted, the length of the airframe is given in feet and inches. Length is measured along the longitudinal axis of the gyroplane. Length is a measurement of the airframe and does not take into account the rotor blades nor extensions such as a pitot tube.

Width Unless otherwise noted, the width is given in feet and inches. The width of the airframe is the airframe's widest point across its lateral axis and does not take into account the rotor blades.

Height Unless otherwise noted, the height is given in feet and inches. Height is the greatest vertical distance of the gyroplane measured from the bottom of its tires (or possibly its pontoons) to its topmost point with the rotor blades installed and assuming the rotor blades are parallel to the length of the fuselage. (This measurement is intended to help determine the vertical clearance required for storing in a hangar or other facility.)

Brakes The kind of brakes is given (e.g., friction or hydraulic). If a parking brake is included, it is noted.

Rotor tachometer If a rotor tachometer is a standard instrument, its type is given (analog or digital).

Engine instruments This is a list of engine instruments included as standard with this model.

Flight instruments This is a list of flight instruments included as standard with this model.

Com/Nav equipment This is a list of electronic communications and navigation devices included as standard equipment with this model.

Optional equipment This is a list of equipment available as options for this model.

Certifications This is a list of any certification(s) (e.g., ASTM, BCAR Section T, etc.) for which this model has been approved and the country/countries in which the certification is approved. This applies only to factory-built models.

Maintenance, warranties, and support policies This is a description of maintenance, warranties and support policies for models.

Available in the US in kit form Models are assumed to be currently available in the US in kit form and can qualify for an Experimental Amateur-Built airworthiness certificate. Some models may be available in another form; e.g., factory-built for certification as Experimental Exhibition or for public use.

FAA approved kit? "Yes" indicates the model is on an FAA list of kits that have been approved for more expeditious certification. A common misunderstanding is that a kit must be on the list to be eligible for certification as Experimental Amateur-Built; however, that is not the case. The list allows for the certification to be expedited to a degree, but the absence of a model on that list does not prevent the model from being certificated as Experimental Amateur Built.

Builder's Assistance? If builder's assistance is available for this model, it is noted.

Notes This is additional information provided by the manufacturer or its agent and may include, for example, notes about specific data items.

Manufacturer: Aerial Surveillance LLC
aka Butterfly Aircraft

Contact information:

Aerial Surveillance LLC
The Butterfly Aircraft
700 Airfield Road
Aurora, Texas 76078
Telephone number: (940) 627-9887
Email: larry@aerialsurveillancellc.com
Web site: www.thebutterflyaircraft.com

US Agents:

The manufacturer's website currently lists 18 "Purchase Contacts" in the US. See the website listed above for that list.

Models

Aerial Surveillance is more popularly known as Butterfly Aircraft. In addition to an ultralight, the manufacturer offers essentially four models of gyroplanes: the single-place Emperor Butterfly, the single-place Monarch Butterfly, the single-place Aurora Butterfly, and the two-place Golden Butterfly. The manufacturer also has produced the "SkyCycle" a unique combination gyroplane/motorcycle.

An Emperor Butterfly

A Monarch Butterfly

An Aurora Butterfly

A Golden Butterfly

No data was provided by the manufacturer.

Manufacturer: Air Command International

Contact information:

Air Command International
P.O. Box 1177
Caddo Mills Airport
Building B
Caddo Mills, Texas 75135
Telephone number: (903) 527-3335
Fax number: (903) 527-3805
Email: aircmd@aircommand.com
Web site: www.aircommand.com

US Agents:

Air Command's website lists two "Dealers"

Britta Penca
85635 East Leewood St
Mammoth, AZ 85618
Phone: (520) 840-0951
Email: Brittaglo@yahoo.com

Ron Menzie
2715 South Main & Airport
Searcy, AR 72143
Phone: (501) 766-6456
ronsgyros.com

Models

Air Command lists a variety of models, several of which apparently vary only according to the engine placed on the airframe. Essentially, the models are these three: a single-place; a two-place with tandem seating; and a two-place with side-by-side seating.

An Air Command single-place model

An Air Command two-place, tandem model

An Air Command two-place, side-by-side model

No data was provided by the manufacturer.

Manufacturer: Autogyro

Contact information:

The US Distributor is
Autogyro USA
Bay Bridge Airport (W29)
210 Airport Rd.
Stevensville, Maryland 21666
Telephone: (410) 604-1719
Email: info@autogyrousa.com
Website: www.AutoGyroUSA.com

US Agents:

The manufacturer, Autogyro Gmbh, is located in Germany. In the US, the manufacturer's distributor is Autogyro USA, whose website lists several dealers in the US. See the website at the address above for a complete listing.

Models

Autogyro offers three models of gyroplanes, all two-place aircraft: the open cockpit, tandem seating MTO Sport; the enclosed cockpit, tandem seating Calidus; and the enclosed cockpit Cavalon with side-by-side seating.

An Autogyro MTO Sport

An Autogyro Calidus

An Autogyro Cavalon

Autogyro Models:	MTO Sport	Calidus	Cavalon
Light-Sport (Yes/No)	Yes	Yes	Yes
First year of production	2001	2008	2010
Number produced	1,300	300	150
Max number of occupants	2	2	2
Seating arrangement	Tandem	Tandem	Side-by-side
Cockpit enclosure	Open	Enclosed	Enclosed
Engine (standard)	Rotax 912ULS	Rotax 912ULS	Rotax 912ULS
Engine (optional) 1	Rotax 914	Rotax 914	Rotax 914
Engine (optional) 2			
Engine (optional) 3			
Engine (optional) 4			
Starter	Electric	Electric	Electric
Rotor blade manufacturer	Autogyro	Autogyro	Autogyro
Rotor blade material	Aluminum	Aluminum	Aluminum
Rotor disk diameter (ft)	27.5	27.5	27.5
Rotor blade cord (in)	8	8	8
Prerotator (kind or none)	Mechanical	Mechanical	Mechanical
Rotor brake (Yes/No)	Yes	Yes	Yes
Propeller	--------------	--------------	--------------
Pusher or tractor	Pusher	Pusher	Pusher
Number of blades	3	3	3
Diameter (in)	68	68	68
Material	CRP	CRP	CRP
Reduction drive (type or none)	Autogyro	Autogyro	Autogyro
Reduction ratio (engine rpm:1)			
Ground adjustable(Yes/No)	Yes	Yes	Yes
Pitch adjustment range			
In-flight adjustable(Yes/No)	Optional	Optional	Optional
Manufacturer	Autogyro	Autogyro	Autogyro
Airframe	--------------	--------------	--------------
Material	Stainless	Stainless	AL/GRP
Construction			
Empty weight (lb)	530	550	590
Maximum takeoff weight (lb)	1,234	1,234	1,234
Useful load (lb)			
Maximum fuel capacity (gal)	18	20	25
Unusable fuel (gal)			

Autogyro Models:	MTO Sport	Calidus	Cavalon
Fuel	Mogas	Mogas	Mogas
Alternate fuel 1	100LL	100LL	100LL
Alternate fuel 2			
Max speed (Vne) (knots)	100	100	87
Max continuous cruise speed (Vh) (knots)	100	100	87
Max range (miles)			
Power for max range (%)			
Max endurance (hrs:mins)	4+ hrs	5 hrs	5+ hrs
Takeoff distance (ft)	300	300	300
Max rate of climb (fpm)			
Landing distance (ft)	0	0	0
Length (ft & in)	16 ft 9 in	15 ft 9 in	15 ft 1in
Width (ft & in)	74.8 in	66.9 in	74.8 in
Height (ft & in)	8 ft 10 in	8 ft 10 in	9 ft 2 in
Brakes			
Parking brake (Yes/No)	Yes	Yes	Yes
Rotor tachometer type (if standard)	Analog	Analog	Analog
Engine instruments standard for this model	--------------	--------------	--------------
Engine instrument 1	Oil pressure	Oil pressure	Oil pressure
Engine instrument 2	Oil temperature	Oil temperature	Oil temperature
Engine instrument 3	Cylinder head temperature	Cylinder head temperature	Cylinder head temperature
Engine instrument 4			
Engine instrument 5			
Engine instrument 6			
Engine instrument 7			
Engine instrument 8			
Flight instruments standard for this model	--------------	--------------	--------------
Flight instrument 1	Airspeed Indicator	Airspeed Indicator	Airspeed Indicator
Flight instrument 2	Altimeter	Altimeter	Altimeter
Flight instrument 3			
Flight instrument 4			
Flight instrument 5			
Flight instrument 6			
Flight instrument 7			
Flight instrument 8			

Autogyro Models:	MTO Sport	Calidus	Cavalon
Com/Nav equipment standard for this model	--------------	--------------	--------------
Com/nav equipment 1	Optional	Optional	Optional
Com/nav equipment 2			
Com/nav equipment 3			
Com/nav equipment 4			
Com/nav equipment 5			
Com/nav equipment 6			
Com/nav equipment 7			
Com/nav equipment 8			
Optional equipment	--------------	--------------	--------------
Optional equipment 1	See website	See website	See website
Optional equipment 2			
Optional equipment 3			
Optional equipment 4			
Optional equipment 5			
Optional equipment 6			
Optional equipment 7			
Optional equipment 8			
Any certifications & country	-------------	--------------	--------------
Certification 1	Just about every country-exception USA	Just about every country-exception USA	Just about every country-exception USA
Certification 1 country			
Certification 2			
Certification 2 country			
Certification 3			
Certification 3 country			
Certification 4			
Certification 4 country			
Maintenance, warranties, and support policies			
Available in US in kit form?	Yes	Yes	Yes
Availability in another form 1			
Availability in another form 2			
Availability in another form 3			
FAA approved kit?	Yes	Yes	Yes
Builder's assistance? (Yes/No)	Yes	Yes	Yes
Notes			

Manufacturer: Aviomania

Contact information:

The US Distributor is
Aviomania USA
4791 73rd St.
La Mesa, California 91942
Telephone: (619) 743-6712
Email: prahq@medt.com
Website: www.AviomaniaUSA.com

US Agent:

The manufacturer, Aviomania Aircraft, is located in Greece. The US representative is Aviomania USA.

Models

Aviomania offers two models of gyroplane: the single-place Genesis G1Sa and the two-place GenesisG2Sa.

The Genesis G1Sa

The Genesis G2Sa

Aviomania Models:	Genesis G1Sa	Genesis G2Sa
Light-Sport (Yes/No)	No	No
First year of production		
Number produced		
Max number of occupants	1	2
Seating arrangement	Single	Tandem
Cockpit enclosure	Pod w/ windshield	Partial
Engine (standard)	Rotax 582	Rotax 912
Engine (optional) 1	Rotax 912	Rotax 914
Engine (optional) 2	MZ 60HP	Hirth 3003, 3702
Engine (optional) 3	Hirth 60-80 HP	Lycoming O320
Engine (optional) 4		
Starter	Optional	Yes
Rotor blade manufacturer	Dragon Wings	Dragon Wings
Rotor blade material	Aluminum	Aluminum
Rotor disk diameter (ft)	23–25	28
Rotor blade cord (in)	7	7
Prerotator (kind or none)	Mechanical or Electric	Mechanical
Rotor brake (Yes/No)	Yes	Yes
Propeller	--------------	--------------
Pusher or tractor	Pusher	Pusher
Number of blades		
Diameter (in)	60–64	70–72
Material	Composite	Varies w/ engine.
Reduction drive (type or none)	Rotax	Varies w/ engine
Reduction ratio (engine rpm:1)	2.64	Varies w/ engine
Ground adjustable (Yes/No)	Yes	Most eng's Yes
Pitch adjustment range		
In-flight adjustable (Yes/No)	Optional	Varies w/ engine
Manufacturer	Ukrainian Hotprop/Warp Drive	Varies w/ engine
Airframe	--------------	--------------
Material	Aluminum & composite	Aluminum & composite
Construction		
Empty weight (lb)	340	530
Maximum takeoff weight (lb)	615	1,330
Useful load (lb)	275	700
Maximum fuel capacity (gal)	7–10	15–19
Unusable fuel (gal)		

Aviomania Models:	Genesis G1Sa	Genesis G2Sa
Fuel	Auto gas	Varies w/ engine
Alternate fuel 1	100LL w/ additive	Auto gas
Alternate fuel 2		100 LL
Max speed (Vne) (knots)	90	95
Max continuous cruise speed (Vh) (knots)	60	70
Max range (miles)	180	260
Power for max range (%)	75	65–70
Max endurance (hrs:mins)	2:30	3:00
Takeoff distance (ft)	150	250
Max rate of climb (fpm)	1,300	1,500 Varies w/ engine
Landing distance (ft)	10	10
Length (ft & in)	12 ft	16 ft
Width (ft & in)	6 ft	7 ft
Height (ft & in)	7 ft 5in	9 ft
Brakes	Mechanical	Mechanical, hydraulic optional
Parking brake (Yes/No)	Optional	Optional
Rotor tachometer type (if standard)	Digital with warning	Digital with warning
Engine instruments standard for this model	--------------	--------------
Engine instrument 1	Dual EGT	2/4 EGT
Engine instrument 2	Water temperature	Water temperature or CHT
Engine instrument 3	RPM	RPM
Engine instrument 4	Hobbs	HOBBS
Engine instrument 5	Fuel level	Fuel level
Engine instrument 6		
Engine instrument 7		
Engine instrument 8		
Flight instruments standard for this model	--------------	--------------
Flight instrument 1	Airspeed indicator	Airspeed indicator
Flight instrument 2	Altimeter	Altimeter
Flight instrument 3	Vertical speed indicator or compass	Vertical speed indicator or compass
Flight instrument 4	Rotor RPM	Rotor RPM
Flight instrument 5		
Flight instrument 6		
Flight instrument 7		
Flight instrument 8		

Aviomania Models:	Genesis G1Sa	Genesis G2Sa
Com/Nav equipment standard for this model	--------------	--------------
Com/nav equipment 1		
Com/nav equipment 2		
Com/nav equipment 3		
Com/nav equipment 4		
Com/nav equipment 5		
Com/nav equipment 6		
Com/nav equipment 7		
Com/nav equipment 8		
Optional equipment	--------------	--------------
Optional equipment 1	EFIS	
Optional equipment 2	Dittel VHF com	
Optional equipment 3	Dittel transponder	
Optional equipment 4	Directional gyro	
Optional equipment 5	Artificial horizon	
Optional equipment 6		
Optional equipment 7		
Optional equipment 8		
Any certifications & country	--------------	--------------
Certification 1	French	French
Certification 1 country		
Certification 2		
Certification 2 country		
Certification 3		
Certification 3 country		
Certification 4		
Certification 4 country		
Maintenance, warranties, and support policies		
Available in US in kit form?	Yes	Yes
Availability in another form 1	Quick build	Quick build
Availability in another form 2	Ready-to-fly	Ready-to-fly
Availability in another form 3		
FAA approved kit?		
Builder's assistance? (Yes/No)	Yes	Yes
Notes		

Manufacturer: Celier Aviation

Contact information:

The US Distributor is
Pictaio Aerospace, LLC
11120 Airport Rd.
Dubuque, Iowa 52003
Telephone: (815) 402-3076
Email: info@PictaioAerospace.com
Website: www.PictaioAerospace.com

US Agent:

In the US, the distributor for Celier Aviation is Pictaio Aerospace in Dubuque, Iowa.

Models

The Celier Aviation two-place, tandem seating Xenon IV is available in the US.

A red Xenon IV

Celier Aviation models:	Xenon IV
Light-Sport (Yes/No)	Yes
First year of production	2013
Number produced	5
Max number of occupants	2
Seating arrangement	Side-by-side
Cockpit enclosure	Enclosed (removable doors)
Engine (standard)	Rotax 912
Engine (optional) 1	Rotax 914
Engine (optional) 2	Rotax 914 (intercooled)
Engine (optional) 3	Rotax 914 (intercooled & upgraded TCU
Engine (optional) 4	
Starter	Electric
Rotor blade manufacturer	Aircopter
Rotor blade material	Aluminum
Rotor disk diameter	Optional: 8.4, 8.6, or 8.8 meters
Rotor blade cord (in)	
Prerotator (kind or none)	Mechanical
Rotor brake (Yes/No)	Yes
Propeller	--------------
Pusher or tractor	Pusher
Number of blades	3
Diameter (in)	68
Material	Composite
Reduction drive (type or none)	Gear
Reduction ratio (engine rpm:1)	2.43
Ground adjustable (yes/no)	Yes
Pitch adjustment range	
In-flight adjustable (yes/no)	Yes
Manufacturer	Kaspar
Airframe	--------------
Material	Composite
Construction	
Empty weight (lb)	640
Maximum takeoff weight (lb)	1,267
Useful load (lb)	627
Maximum fuel capacity (gal)	22
Unusable fuel (gal)	0.25

Celier Aviation models:	Xenon IV
Fuel	Unleaded
Alternate fuel 1	100LL
Alternate fuel 2	
Max speed (Vne) (knots)	113
Max continuous cruise speed (Vh) (knots)	113
Max range (miles)	400
Power for max range (%)	
Max endurance (hrs & mins)	4–5 hrs
Takeoff distance (ft)	150–300
Max rate of climb (fpm)	
Landing distance (ft)	0–30
Length (ft & in)	16 ft 6 in
Width (ft & in)	7 ft 5 in
Height (ft & in)	9 ft 0 in
Brakes	Hydraulic
Parking brake (Yes/No)	Yes
Rotor tachometer type (if standard)	Electric
Engine instruments standard for this model	--------------
Engine instrument 1	Manifold pressure (if tubo)
Engine instrument 2	Fuel level
Engine instrument 3	Rotor speed
Engine instrument 4	Digital instruments parameter display
Engine instrument 5	
Engine instrument 6	
Engine instrument 7	
Engine instrument 8	
Flight instruments standard for this model	--------------
Flight instrument 1	Airspeed indicator
Flight instrument 2	Vertical speed indicator
Flight instrument 3	Altimeter
Flight instrument 4	Skid indicator
Flight instrument 5	Ball compass
Flight instrument 6	
Flight instrument 7	
Flight instrument 8	

Celier Aviation models:	Xenon IV
Com/Nav equipment standard for this model	--------------
Com/nav equipment 1	
Com/nav equipment 2	
Com/nav equipment 3	
Com/nav equipment 4	
Optional equipment	--------------
Optional equipment 1	Carbon fiber body
Optional equipment 2	Sound proofing
Optional equipment 3	Cabin heat/electric heated seats
Optional equipment 4	Electric rotor trim
Optional equipment 5	Leather seating
Optional equipment 6	Wheel pants
Optional equipment 7	Amphibious floats
Optional equipment 8	VHF radio/transponder
Optional equipment 9	LED lighting
Optional equipment 10	Larger rotor
Optional equipment 11	Door locks
Optional equipment 12	Upgraded avionics/glass cockpit
Any certifications & country	-------------
Certification 1	
Certification 1 country	
Certification 2	
Certification 2 country	
Certification 3	
Certification 3 country	
Certification 4	
Certification 4 country	
Maintenance, warranties, and support policies	12 month warranty
Available in US in kit form?	Yes
Availability in another form 1	Fully assembled for "public use" aircraft
Availability in another form 2	
Availability in another form 3	
FAA approved kit?	In process
Builder's assistance? (Yes/No)	Yes
Notes	

Manufacturer: Halley Ltd.

Contact information:

The US Distributor is
SilverLight Aviation
40417 Chancey Rd. Suite 101
Zephyrhills, Florida 33542
Telephone: (813) 786-8290
Email: info@SilverLightAviation.com
Website: www.SilverLightAviation.com

US Agent:

The manufacturer, Halley Ltd., is located in Hungary. The US distributor is SliverLight Aviation in Zephryhills, Florida.

Models

Halley Ltd., a well-established manufacturer of small airplanes and trikes, offers the Apollo AG-1 as its entry into the gyroplane market. The AG-1's cockpit is modeled after the Apollo line of trikes.

An Apollo AG-1 (above and below)

Halley Models:	Apollo AG-1
Light-Sport (Yes/No)	Yes
First year of production	2011
Number produced	27
Max number of occupants	2
Seating arrangement	Tandem
Cockpit enclosure	Semi enclosed
Engine (standard)	Rotax 912ULS – 100 HP
Engine (optional) 1	914UL – 115 HP Turbo
Engine (optional) 2	
Engine (optional) 3	
Engine (optional) 4	
Starter	Electric
Rotor blade manufacturer	Averso
Rotor blade material	Aluminum alloy
Rotor disk diameter (ft)	27.6
Rotor blade cord (in)	8.4
Prerotator (kind or none)	Pneumatic
Rotor brake (Yes/No)	Yes
Propeller	Aero Prop or Sterna
Pusher or tractor	Pusher
Number of blades	3
Diameter (in)	69
Material	Composite
Reduction drive (type or none)	Rotax integrated gearbox
Reduction ratio (engine rpm:1)	2.43
Ground adjustable (yes/no)	Yes
Pitch adjustment range	
In-flight adjustable (yes/no)	No
Manufacturer	Aero Prop or Sterna
Airframe	--------------
Material	Stainless steel alloy
Construction	Tig welded frame with composite fairings
Empty weight (lb)	630
Maximum takeoff weight (lb)	1,130
Useful load (lb)	500
Maximum fuel capacity (gal)	16.5
Unusable fuel (gal)	

Halley Models:	Apollo AG-1
Fuel	93 octane Mogas
Alternate fuel 1	100LL Avgas
Alternate fuel 2	93 octane E10
Max speed (Vne) (knots)	100
Max continuous cruise speed (Vh) (knots)	92
Max range (miles)	333
Power for max range (%)	55
Max endurance (hrs & mins)	4:45
Takeoff distance (ft)	300 (912ULS) / 200 (914UL)
Max rate of climb (fpm)	780 (912ULS) / 984 (914 UL)
Landing distance (ft)	0–30
Length (ft & in)	
Width (ft & in)	
Height (ft & in)	
Brakes	Hydraulic (DOT 4)
Parking brake (Yes/No)	Yes
Rotor tachometer type (if standard)	Analog
Engine instruments standard for this model	--------------
Engine instrument 1	CHT
Engine instrument 2	RPM (tachometer)
Engine instrument 3	Oil pressure gauge
Engine instrument 4	Oil temperature gauge
Engine instrument 5	
Engine instrument 6	
Engine instrument 7	
Engine instrument 8	
Flight instruments standard for this model	--------------
Flight instrument 1	Air speed indicator
Flight instrument 2	Altimeter
Flight instrument 3	Vertical speed indicator
Flight instrument 4	
Flight instrument 5	
Flight instrument 6	
Flight instrument 7	
Flight instrument 8	

Halley Models:	Apollo AG-1
Com/Nav equipment standard for this model	--------------
Com/nav equipment 1	
Com/nav equipment 2	
Com/nav equipment 3	
Com/nav equipment 4	
Com/nav equipment 5	
Com/nav equipment 6	
Com/nav equipment 7	
Com/nav equipment 8	
Optional equipment	--------------
Optional equipment 1	MGL V6 Com
Optional equipment 2	Trig Mode S transponder or Sandia Mode C transponder
Optional equipment 3	Instructor back seat analog airspeed indicator and altimeter
Optional equipment 4	
Optional equipment 5	
Optional equipment 6	
Optional equipment 7	
Optional equipment 8	
Any certifications & country	-------------
Certification 1	ULM
Certification 1 country	Hungary
Certification 2	
Certification 2 country	
Certification 3	
Certification 3 country	
Certification 4	
Certification 4 country	
Maintenance, warranties, and support policies	Builder's assist in the USA, 2 year warranty on materials and workmanship on airframe. Engine and avionics warranties from equipment manufacturers
Available in US in kit form?	Yes
Availability in another form 1	
Availability in another form 2	
Availability in another form 3	
FAA approved kit?	In works
Builder's assistance? (Yes/No)	Yes
Notes	

Manufacturer: Magni Gyro

Contact information:

The US Distributor is
Magni USA, LLC
17225 Pleasant View Dr.
Ste. Genevieve, Missouri 63670
Telephone: (314) 540-0367
Fax number: (573) 883-2866
Email: gyrogreg@ldd.net
Website: www.MagniGyro.com

US Agents:

The manufacturer, Magni Gyro Srl, is located in Italy. In the US, the manufacturer's distributor is Magni USA, LLC (see contact information above).

Other agents for Magni Gyro in the US are

Lone Star Magni Gyro, Inc.
Taylor Municipal Airport (J74)
Taylor, Texas 76574
Telephone: (800) 241-0924
Email: info@LoneStarMagniGyro.com
Website: www.LoneStarMagniGyro.com

MagniFlight LLC
Email: MagniFlight1@charter.net
Website: www.MagniFlightLLC.com

Models

Magni offers three models of gyroplanes in the US, all two-place aircraft: the open cockpit, tandem seating M-16 Trainer with dual controls; the open cockpit, tandem seating M-22 Voyager with optional rear-seat controls and with luggage pods; and the enclosed cockpit M-24 Orion with side-by-side seating.

A Magni M-16 Trainer

A Magni M-22 Voyager

A Magni M-24 Orion

Magni Models:	M-16	M-22	M-24
Light-Sport (Yes/No)	Yes	Yes	Yes
First year of production	1992	2005	2008
Number produced	450	180	150
Max number of occupants	2	2	2
Seating arrangement	Tandem	Tandem	Side-by-side
Cockpit enclosure	Open	Open	Enclosed
Engine (standard)	Rotax 912UL	Rotax 914UL	Rotax 914UL
Engine (optional) 1	Rotax 914ULS		
Engine (optional) 2	Rotax 912iS		
Engine (optional) 3			
Engine (optional) 4			
Starter	Yes	Yes	Yes
Rotor blade manufacturer	Magni Gyro SRL	Magni Gyro SRL	Magni Gyro SRL
Rotor blade material	Composite	Composite	Composite
Rotor disk diameter (ft)	28	28	28
Rotor blade cord (in)	8	8	8
Prerotator (kind or none)	Flex Cable/Bendix	Flex Cable/Bendix	Flex Cable/Bendix
Rotor brake (Yes/No)	Yes	Yes	Yes
Propeller	--------------	--------------	--------------
Pusher or tractor	Pusher	Pusher	Pusher
Number of blades	3	3	3
Diameter (in)	68	68	68
Material	Composite	Composite	Composite
Reduction drive (type or none)	Gear	Gear	Gear
Reduction ratio (engine rpm:1)	2.43	2.43	2.43
Ground adjustable(Yes/No)	Yes	Yes	Yes
Pitch adjustment range	Unlimited	Unlimited	Unlimited
In-flight adjustable(Yes/No)	Optional	Optional	Optional
Manufacturer	Arplast Héliec	Arplast Héliec	Arplast Héliec
Airframe	--------------	--------------	--------------
Material	4130 steel	4130 steel	4130 steel
Construction	Welded	Welded	Welded
Empty weight (lb)*	585	595	629
Max takeoff weight (lb)*	1,212	1,212	1,212
Useful load (lb)*	627	617	583
Maximum fuel capacity (gal)	19	19	21
Unusable fuel (gal)	0.5	0.5	0.5

Magni Models:	M-16	M-22	M-24
Fuel	91 octane 10% Ethanol max	91 octane 10% Ethanol max	91 octane 10% Ethanol max
Alternate fuel 1	Avgas 100LL	Avgas 100LL	Avgas 100LL
Max speed (Vne) (knots)	100 (115mph)	100 (115mph)	91 (105 mph)
Max continuous cruise speed (Vh) (knots)	78 (90 mph)	78 (90 mph)	74 (85 mph)
Max range (miles)	300	300	300
Power for max range (%)	80	80	90
Max endurance (hrs:mins)**	3:24	3:24	3:30
Takeoff distance (ft)	230	230	245
Max rate of climb (fpm)	950	950	950
Landing distance (ft)	0–100	0–100	0–100
Length (ft & in)	15 ft 4 in	15 ft 4 in	15 ft 4 in
Width (ft & in)	5 ft 11 in	5 ft 11 in	5 ft 11 in
Height (ft & in)	8 ft 7 in	8 ft 7 in	8 ft 7 in
Brakes	Hydraulic	Hydraulic	Hydraulic
Parking brake (Yes/No)	Optional-Standard for C version	Optional	Optional-Standard for C version
Rotor tachometer type (if standard)	Digital	Digital	Digital
Engine instruments standard for this model	Digital Engine Monitor – includes:	Digital Engine Monitor – includes:	Digital Engine Monitor – includes:
Engine instrument 1	RPM	RPM	RPM
Engine instrument 2	Oil temp & press	Oil temp & press	Oil temp & press
Engine instrument 3	Fuel pressure	Fuel pressure	Fuel pressure
Engine instrument 4	Battery voltage	Battery voltage	Battery voltage
Engine instrument 5	4 EGT	4 EGT	4 EGT
Engine instrument 6	2 CHT	2 CHT	2 CHT
Engine instrument 7	2 CHT	2 CHT	2 CHT
Engine instrument 8	Flt & Hobbs timer	Flt & Hobbs timer	Flt & Hobbs timer
Flight instruments standard for this model	--------------	--------------	--------------
Flight instrument 1	Altimeter	Altimeter	Altimeter
Flight instrument 2	Airspeed indicator	Airspeed indicator	Airspeed indicator
Flight instrument 3	Magnetic compass (optional vertical card type)	Magnetic compass (optional vertical card type)	Vertical card magnetic compass
Flight instrument 4	Fuel gauge	Fuel gauge	Fuel gauge
Com/Nav equipment standard for this model	--------------	--------------	--------------
Com/nav equipment 1	Com radio & intercom	Com radio & intercom	Com radio & intercom

Magni Models:	M-16	M-22	M-24
Optional equipment	--------------	--------------	--------------
Optional equipment 1	Manifold pressure	Manifold pressure	Manifold pressure
Optional equipment 2	Fuel flow	Fuel flow	Fuel flow
Optional equipment 3	Transponder	Transponder	Transponder
Optional equipment 4	Rear baggage compartment		
Optional equipment 5	Front baggage compartments		
Optional equipment 6	GPS/EFIS	GPS/EFIS	GPS/EFIS
Optional equipment 7	Custom paint	Custom paint	Custom paint
Optional equipment 8	Panel power outlet	Panel power outlet	Panel power outlet
Optional equipment 9	Spinner	Spinner	Spinner
Optional equipment 10	Step-on	Step-on	
Optional equipment 11			Mast fairing
Optional equipment 12			"VIP" finishing
Certifications & country***	-------------	--------------	--------------
Certification 1	Section T		Section T
Certification 1 country	UK (C version)		UK (C version)
Certification 2	BUT	BUT	BUT
Certification 2 country	Germany	Germany	Germany
Certification 3			
Certification 3 country	Spain	Spain	Spain
Certification 4			
Certification 4 country	Canada		
Certification 5			
Certification 5 country	RSA	RSA	RSA
Maintenance, warranties, and support policies	See Notes****	See Notes****	See Notes****
Available in US in kit form?	Yes (E-AB, 51%)	Yes (E-AB, 51%)	Yes (E-AB, 51%)
Availability in another form1	Fully-built (E-Exhibition)	Fully-built (E-Exhibition)	Fully-built (E-Exhibition)
Availability in another form2	Fully built (public use)	Fully built (public use)	Fully built (public use)
FAA approved kit?	Yes		
Builder's assistance? (Yes/No)	Yes	Yes	Yes

Notes: *Weights are US. Weight depends on national rules & limitations. Check with factory or local agent.

**Endurance are US numbers using available 10% Ethanol at 6 gal/hr. 4 hr endurance if using no Ethanol at 5.5 gal/hr. In Europe, we consider an average of 4 hr. endurance.

*** The certifications apply to factory-built gyroplanes.

****2 year limited warranty on airframe, builder support, owner maintenance support. For factory-built, 2 years warranty. Standard Rotax warranty (100 hours) on engine.

Manufacturer: Rotor Flight Dynamics, Inc.

Contact information:

Rotor Flight Dynamics, Inc.
19242 Grange Hall Loop
Wimauma, Florida 33598
Telephone number: (813) 634-3370
Fax number: (813) 634-3370
Email: rfdlouie@hotmail.com
Web site: www.rotorflightdynamicsinc.com

US Agents:

The manufacturer's website doesn't list any agents.

Models

Rotor Flight Dynamics manufacturers the "Dragon Wings" rotor blades. In addition to an ultralight, the manufacturer offers two models of gyroplanes: the single-place Dominator and the two-place, tandem seating Dominator.

An single-place Dominator

A two-place Dominator

No data was provided by the manufacturer.

Manufacturer: Sport Copter, Inc.

Contact information:

Sport Copter, Inc.
34012 Skyway Dr.
Scappoose, Oregon 97056
Telephone number: (503) 543-7000
Fax number: (503) 543-7041
Email: information@sportcopter.com
Web site: www.SportCopter.com

US Agents:

The manufacturer's website doesn't list any agents.

Models

Sport Copter offers two models of gyroplanes that fall within the definition of a light-sport aircraft: the Vortex and the Vortex M12. Both are single-place aircraft.

Sport Copter has also developed the Sport Copter II, a two-place, enclosed gyroplane with side-by-side seating. Its gross weight places it beyond the maximum allowed takeoff weight allowed for light-sport aircraft. The manufacturer's website suggests that the company intends to develop an "LSA version" of the Sport Copter II whenever the FAA allows factory-built SLSA/ELSA gyroplanes.

A Sport Copter Vortex

A Sport Copter Vortex M12

A Sport Copter II

Sport Copter Models:	Vortex	Vortex M912	Sport Copter II
Light-Sport (Yes/No)	Yes	Yes	Version in Process
First year of production	1995		2007 (see Notes below)
Number produced			
Max number of occupants	1	1	2
Seating arrangement			Side-by-side
Cockpit enclosure	Partial	Partial	Fully enclosed
Engine (standard)	Rotax 582 C drive	Rotax 912ULS	Lycoming IO-360
Engine (optional) 1	Rotax 582 E drive		
Engine (optional) 2			
Engine (optional) 3			
Engine (optional) 4			
Starter	Pull or Electric	Electric	Electric
Rotor blade manufacturer	Sport Copter	Sport Copter	Sport Copter
Rotor blade material	Aluminum	Aluminum	Aluminum
Rotor disk diameter (ft)	25	27	31
Rotor blade cord (in)	8	8	9
Prerotator (kind or none)	Mechanical	Mechanical	Mechanical
Rotor brake (Yes/No)	Yes	Yes	Yes
Propeller	--------------	--------------	--------------
Pusher or tractor	Pusher	Pusher	Pusher
Number of blades	3	3	3
Diameter (in)	66	68	74
Material	Composite	Composite	Composite
Reduction drive (type or none)	Gearbox	Gearbox	
Reduction ratio (engine rpm:1)	3:1	2.43:1	
Ground adjustable(Yes/No)	Yes	Yes	Cockpit
Pitch adjustment range			
In-flight adjustable(Yes/No)	No	No	Controllable
Manufacturer	Warp Drive	Warp Drive	MT Propeller
Airframe	--------------	--------------	--------------
Material	6061 Alum	4130 and 6061	4130/6061/Carbon Fiber
Construction	Welded/bolted	Welded/bolted	Welded/bolted/ composite
Empty weight (lb)	420	565	1,000
Maximum takeoff weight (lb)	760	983	1,700
Useful load (lb)			
Maximum fuel capacity (gal)	14.5	18.5	34
Unusable fuel (gal)			

Sport Copter Models:	Vortex	Vortex M912	Sport Copter II
Fuel	92 auto fuel (US)	92 auto fuel (US)	100LL Avgas
Alternate fuel 1	95 auto fuel (non US)	95 auto fuel (non US)	
Alternate fuel 2	Avgas	Avgas	
Max speed (Vne) (* mph)	80 mph*	120 mph*	120 mph*
Max continuous cruise speed (Vh) (*mph)	70-75 mph*	110 mph*	110 mph*
Max range (miles)	170	240+	300
Power for max range (%)			
Max endurance (hrs:mins)			
Takeoff distance (ft)	10–80	10–50	50–200
Max rate of climb (fpm)	600–800	700–1,000	1,000
Landing distance (ft)	0–20	0–20	0–20
Length (ft & in)	160 in	160 in	186 in
Width (ft & in)	76 in	76 in	81 in
Height (ft & in)	108.5 in	108.5 in	119 in
Brakes			
Parking brake (Yes/No)	Optional	Optional	optional
Rotor tachometer type (if standard)	Digital	Digital	Digital
Engine instruments standard for this model	--------------	--------------	--------------
Engine instrument 1	Engine RPM	Engine RPM	Dynon EMS D10
Engine instrument 2	Water temperature	Water temperature	
Engine instrument 3	EGT	EGT	
Engine instrument 4	EGT	EGT	
Engine instrument 5		Oil pressure	
Engine instrument 6		Oil temperature	
Engine instrument 7		Voltmeter	
Engine instrument 8			
Flight instruments standard for this model	--------------	--------------	--------------
Flight instrument 1	Airspeed indicator	Airspeed indicator	Dynon EFIS D100
Flight instrument 2	Vertical speed indicator	Vertical speed indicator	Vertical card compass
Flight instrument 3	Altimeter	Altimeter	
Flight instrument 4	Hobbs hour meter	Hobbs hour meter	
Flight instrument 5			
Flight instrument 6			
Flight instrument 7			
Flight instrument 8			

Sport Copter Models:	**Vortex**	**Vortex M912**	**Sport Copter II**
Com/Nav equipment standard for this model	--------------	--------------	--------------
Com/nav equipment 1			Garmin SL40
Com/nav equipment 2			Garmin GMA240
Com/nav equipment 3			
Com/nav equipment 4			
Com/nav equipment 5			
Com/nav equipment 6			
Optional equipment	--------------	--------------	--------------
Optional equipment 1	MicroAir Transceiver	MicroAir Transceiver	Dynon Skyview
Optional equipment 2	LED Nav/Pos Strobes	LED Nav/Pos Strobes	MGL Xtreme EMS
Optional equipment 3	4-way Trim	4-way Trim	Garmin GTX327
Optional equipment 4	Vertical Card Compass	Vertical Card Compass	Garmin Aeroa 560
Optional equipment 5	MGL Xtreme EMS	MGL Xtreme EMS	ICOM IC-A210 transceiver
Optional equipment 6			Garmin GTN750
Optional equipment 7			
Optional equipment 8			
Any certifications & country	-------------	--------------	--------------
Certification 1		Australia	
Certification 1 country			
Certification 2			
Certification 2 country			
Maintenance, warranties, and support policies			
Available in US in kit form?	Yes	Yes	In process
Availability in another form 1			
Availability in another form 2			
Availability in another form 3			
FAA approved kit?	In process	In process	In process
Builder's assistance? (Yes/No)	Yes	Yes	Yes
Additional information			
Notes			First one produced in 2007. We halted the production for an engine change. Switched from Subaru to Lycoming.

20742486R00071

Made in the USA
San Bernardino, CA
21 April 2015